Thirteen Papers in Algebra

AMERICAN MATHEMATICAL SOCIETY
TRANSLATIONS

Series 2

Volume 132

Thirteen Papers in Algebra

by

I. G. Dmitriev
S. Kh. Darbinyan
A. V. Kostochka
O. G. Rudenskaya
T. M. Zuparov
V. V. Iliev
D. A. Leĭtes
V. Ya. Golodets
N. I. Nessonov
I. S. Bezverkhnyaya
V. A. Grinblat
N. A. Vavilov
A. L. Vishnevetskiĭ

AMERICAN MATHEMATICAL SOCIETY
PROVIDENCE, RHODE ISLAND

Translation edited by BEN SILVER

1980 *Mathematics Subject Classification* (1985 *Revision*). Primary 05CXX, 11K50, 17A70, 20E06, 20FXX, 20G35, 46L35; Secondary 17B70, 46L40, 60F05, 81C40.

Library of Congress Cataloging-in-Publication Data
Thirteen papers in algebra.

(American Mathematical Society translations, ISSN 0065-9290; ser. 2, v. 132)
1. Algebra. I. Dmitriev, I. G. II. Series.
QA3.A572 ser. 2, vol. 132 510 s [512] 86-17254
[QA155.2]
ISBN 0-8218-3107-0

Contents

Russian Contents*

И. Г. Дмитриев, Характеризация класса K-деревьев, Методы дискретного анализа, № 38 (1982), стр. 9–18

С. Х. Дарбинян, Контуры любой длины в орграфах с большими полустепенями, Доклады Акад. Наук Армян. ССР, т. 75 (1982), № 4, стр. 147–152

А. В. Косточка, О минимуме числа Хадвигера для графов с данной средней степенью вершин, Методы дискретного анализа, № 38 (1982), стр. 37–58

О. Г. Руденская, Верхняя оценка числа ребер гиперграфа с данным диаметром, Математич. исследования № 66 (1982), стр. 141–147

Т. М. Зупаров, Об одной теореме из метрической теории непрерывных дробей, Известия Акад. Наук Узбек. ССР, Серия физ.-матем. наук, 1981, № 6, стр. 9–12

В. В. Илиев, Поверхности с $p_g = 3, K^2 = 3$, Часть I, Сердика. Българско мат. списание, т. 6 (1980), № 4, 352–362

В. В. Илиев, Поверхности с $p_g = 3, K^2 = 3$, Часть II, Сердика. Българско мат. списание, т. 7 (1981), № 4, 390–395

Д. А. Лейтес, Формула для характеров неприводимых конечномерных представлений супералгебр Ли серии C, Доклады Болгарск. Акад. Наук, т. 33 (1980), № 8, стр. 1053–1055

В. Я. Голодец и Н. И. Нессонов, Свободные группы, T-свойство, II_1-факторы с различными счетными фундаментальными группами, Доклады Акад. Наук Украин. ССР, Серия A, 1983, № 8, стр. 7–10

*The American Mathematical Society scheme for transliteration of Cyrillic may be found at the end of index issues of *Mathematical Reviews*.

И. С. Безверхняя, О сопряженности конечных множеств подгрупп в свободном произведении групп, Алгоритмич. проблемы теории групп и полугрупп, Тульский Гос. Педагогич. Ин-т, Тула, 1981, стр. 102–116

В. А. Гринблат, О нормализаторах групп Артина, Алгоритмич. проблемы теории групп и полугрупп, Тульский Гос. Педагогич. Ин-т, Тула, 1981, стр. 82–94

А. Л. Вишневецкий, Группы класса 2 и экспоненты p с коммутантом порядка p^2, Доклады Акад. Наук Украин. ССР, Серия А, 1980, № 9, стр. 9–11

Н. А. Вавилов, Подгруппы полной линейной группы над кольцом, содержащие группу клеточно-треугольных матриц, Вестник Ленинград. Унив., 1977, № 19, стр. 139–140

Amer. Math. Soc. Transl.
(2) Vol. **132**, 1986

Characterization of the Class of k-Trees

UDC 519.1

I. G. DMITRIEV

We shall consider only ordinary graphs. All the concepts and notation not defined here can be found in [1] or [2]. A graph G with $p \geq k+1$ vertices is called a *k-tree* ($k \geq 0$) if it either coincides with a $(k+1)$-vertex complete graph or is obtained from a $(p-1)$-vertex k-tree by adding a new vertex, adjacent to each vertex of a certain subgraph with k vertices. In what follows the class of k-trees will be denoted by T^k.

Harary and Palmer [3] defined the concept of a k-plex as an n-dimensional simplicial complex in which every r-simplex ($r \leq k$) is contained in some k-simplex. Bieneke and Pippert [4] gave a characterization of k-trees in the notions of k-plex. Rose [5] gave a characterization of k-trees in terms of cliques, chains and separating sets.

In this paper we give a characterization of k-trees by their chromatic polynomials.

THEOREM 1. *A graph G with $p \geq k+1$ vertices is a k-tree if and only if its chromatic polynomial $f(G,\lambda) = \lambda(\lambda-1)\cdots(\lambda-k+1)(\lambda-k)^{p-k}$, where $k \geq 0$.*

Let us give the necessary definitions and notation. Let U^{k+1} denote the class of uniquely $(k+1)$-colorable graphs G with $|e(G)| = pk - \binom{k+1}{2}$ edges, where $p \geq k+1$ is the number of vertices of G. From the definition of a k-tree it follows that $T^k \subset U^{k+1}$. A graph is said to be *strongly cyclic* (or a *graph without holes*) if any simple cycle of it of length $l \geq 4$ has a chord. Otherwise it is said to be *weakly cyclic*. Obviously, all k-trees are strongly cyclic. If we are given a graph $G \in U^{k+1}$ and a coloring of it, then the monochromatic classes are denoted by $V_1, \ldots, V_{k+1}$, and the cardinality of V_i by p_i. A subgraph generated by the notion

1980 *Mathematics Subject Classification* (1985 *Revision*). Primary 05C05; Secondary 05C15, 05C38.

Translation of Metody Diskret. Analiz. Vyp. 38 (1982), 9–18; MR **85c:** 05016.

of m monochromatic classes is called an *m-color subgraph*. The *encirclement* of a vertex x is understood to be the subgraph generated by vertices adjacent to x. We shall say that two vertices are *monochromatic* if under any minimal coloring they have the same color. A *k-clique* of a graph will be understood as a k-vertex complete subgraph of it, not necessarily maximal.

For the proof of the main result we need the following assertions.

LEMMA 1. *In a graph $G \in U^{k+1}$ any 2-color subgraph is a tree.*

PROOF. Obviously, in view of the unique colorability of G, all 2-color subgraphs are connected ([2], Theorem 12.16). Assuming the contrary, we obtain

$$|e(G)| > \sum_{1 \le i < j \le k+1} (p_i + p_j - 1) = pk - \binom{k+1}{2},$$

which contradicts the definition of the class U^{k+1}. This proves Lemma 1.

LEMMA 2. *The encirclements of any pair of monochromatic vertices of a graph G of U^{k+1} contain no more than k common vertices.*

PROOF. Assuming the contrary, we find that two monochromatic vertices are simultaneously adjacent to two other vertices of the same color; that is, there is a 2-color cycle in G. This contradicts Lemma 1, so Lemma 2 is proved.

LEMMA 3. *A graph $G \in U^{k+1} \backslash T^k$ is weakly cyclic for any value of $k \ge 2$.*

PROOF. Assume the contrary. Then, as Dirac showed [6], there is a simplicial vertex in G (the encirclement of which is a complete subgraph). Since G is uniquely colorable and $(k+1)$-chromatic, this vertex has degree k. From this by simple arguments we see that G is a k-tree. This is a contradiction, so Lemma 3 is proved.

LEMMA 4. *Suppose that $G \in T^k$ and has $p \ge k+2$ vertices ($k \ge 1$). Then in G there is a pair of vertices x and y whose encirclements intersect in a k-clique, and this k-clique is a minimal separating set of the vertices x and y.*

PROOF. The existence of such vertices follows immediately from the definition of a k-tree. The proof of the second part can be found in [5].

From now on vertices x and y that are adjacent to a common k-clique Q_k will be called the *ends of the hyperedge* $[xQ_ky]$ of order k.

LEMMA 5. *Suppose that $G \in T^k$ and that $[xQ_ky]$ is some hyperedge of it. Then the graph G' obtained by identifying the vertices x and y is a k-tree.*

PROOF. We observe that the number of edges is

$$|e(G')| = |e(G)| - k = (p-1) \cdot k - \binom{k+1}{2}$$

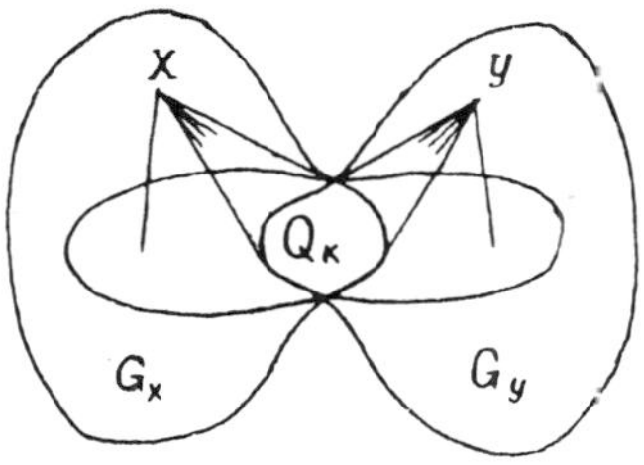

FIGURE 1

and G' is uniquely $(k+1)$-colorable; that is, $G' \in U^{k+1}$. Therefore, by Lemma 3, it remains to show that G' is strongly cyclic. Suppose that G' has a cycle C_l of length $l \geq 4$ without chords. By Lemma 4, G is the union of two of its subgraphs that intersect in a k-clique Q_k (see Figure 1). We denote them by G_x and G_y. Obviously, the graph G' obtained after identifying x and y is also the union of two subgraphs, namely the subgraphs that coincide with G_x and G_y and intersect in the $(k+1)$-clique $Q_k + \{x \equiv y\}$. Then, since this $(k+1)$-clique is complete, the cycle C_l in the graph G' belongs to one of the subgraphs G_x or G_y, which contradicts the fact that the k-tree G is strongly cyclic. This proves Lemma 5.

LEMMA 6. *Suppose that $G \in T^2$ and that $P(x, y)$ is a 2-color chain of it of length no less than 2. Then a vertex z of a third color, adjacent to the ends of the chain $P(x, y)$, is also adjacent to all its internal vertices.*

PROOF. By Lemma 1, all 2-color subgraphs are acyclic. Therefore, the chain $P(x, y)$ has no chord. Assuming the contrary, we obtain a 3-color cycle without chords and of length $l \geq 4$, which contradicts the fact that the 2-tree G is strongly cyclic. This proves Lemma 6.

THEOREM 2. *A graph $G \in U^3$ is a 2-tree if and only if the number of triangles $t(G)$ (cycles of length 3) of it is given by $t(G) = p - 2$, where p is the number of vertices of G.*

PROOF. *Necessity* follows from the definition of a 2-tree.

Sufficiency. For this we show that in a p-vertex graph $G \in U^3 \backslash T^2$ the number $t(G) < p - 2$. Assume the contrary. Suppose that there is a graph $G \in U^3 \backslash T^2$, minimal in the number of vertices, for which the number of triangles $t(G) \geq p - 2$. There are no vertices of degree 2 in G. Otherwise, if a vertex of degree 2 were removed, we should obtain a graph $G' = G - x$ with $p - 1$ vertices that belongs to the class U^3. If the vertex x is adjacent to an edge, we have $t(G') = t(G) - 1 \geq (p-1) - 2$; otherwise $t(G') = t(G) > (p-1) - 2$. In the first case, since G is minimal the graph G' is a 2-tree, but then G is also a 2-tree. This contradicts the fact that $G \in U^3 \backslash T^2$. In the second case we immediately obtain a contradiction with the fact that G is minimal. In G there is a hyperedge $[xQ_2y]$. Otherwise every edge would belong to at most one triangle; that is, $2p - 3 \geq 3t(G) \geq 3(p-2)$, or $p \leq 3$. This contradicts the fact that $G \in U^3 \backslash T^2$.

Now, identifying the monochromatic vertices x and y, we obviously obtain a $(p-1)$-vertex graph $G' \in U^3$. By Lemma 2, $t(G') = t(G) - 1 \geq (p-1) - 2$. If G' is not a 2-tree, we obtain a contradiction with the fact that G is minimal.

Suppose that G' is a 2-tree. Since the vertices of G have degree not less than 3, after the identification only the degrees of the vertices of the 2-clique Q_2 can be equal to 2, and this contradicts the fact that in any 2-tree with $p \geq 4$ vertices there are no fewer than two vertices of degree 2, and they are not adjacent. This proves Theorem 2. From the proof of this theorem there immediately follows the next result.

COROLLARY 2.1. *If a p-vertex graph $G \in U^3 \backslash T^2$, then $t(G) < p - 2$.*

THEOREM 3. *For $k \geq 3$ a $(k+1)$-colorable graph is a k-tree if and only if in some $(k+1)$-coloring of it every k-color subgraph is a $(k-1)$-tree.*

PROOF. *Necessity* is obvious.

Sufficiency. Suppose that the theorem is false, starting with some $k \geq 3$, and that G is minimal in the number of vertices and satisfies the conditions of Theorem 3, but is not a k-tree. By the condition of the theorem and Lemma 1, all 2-color subgraphs are trees. Therefore, the number of edges $|e(G)| = pk - \binom{k+1}{2}$, where p is the number of vertices of G.

We show that there is a hyperedge of order k in G. By Lemma 4, in a $(k-1)$-tree, say $G[V \backslash V_1]$, there is a hyperedge $[sQ_{k-1}t]$. Otherwise, all the $(k-1)$-trees $G[V \backslash V_i]$ $(i = 1, \dots, k+1)$ would have k vertices; that is, G would be a $(k+1)$-vertex complete graph, which contradicts our assumption. Without loss of generality we may assume that the monochromatic vertices s and t have the color 2. By Lemma 1, between s and t there is a unique 2-color chain $P(s,t)$ of colors 1 and 2 joining them, and obviously of length not less than 2. By Lemma 6, all the vertices of the $(k-1)$-clique Q_{k-1} are adjacent to all the vertices of the chain $P(s,t)$, since all the colors $i = 3, \dots, k+1$ occur in Q_{k-1}, and obviously all the 3-color subgraphs $G[V_1 \cup V_2 \cup V_i]$ are 2-trees.

Then, taking for x and y two arbitrary vertices of the chain $P(s,t)$ at a distance 2, we obtain a hyperedge $[xQ_ky]$ of order k. Without loss of generality we may assume that the monochromatic vertices x and y have color 1. The fact that they are monochromatic follows from the fact that the $(k+1)$-colorable graph G has a $(k+1)$-clique and the vertices x and y are the ends of a hyperedge of order k. Under the identification of x and y the $(k-1)$-tree $G[V \backslash V_1]$ goes into itself, and by Lemma 5 the remaining $(k-1)$-trees $G[V \backslash V_i]$, $i = 2, \dots, k+1$, go into $(k-1)$-trees $G[V' \backslash V_i]$. Then, by induction, the graph $G' = G[V']$ is a k-tree. Since x and y are monochromatic, any $(k+1)$-colorable graph G determines a certain $(k+1)$-coloring of G', and conversely.

Thus, a p-vertex graph G with $pk - \binom{k+1}{2}$ edges is uniquely $(k+1)$-colorable; that is, $G \in U^{k+1}$. Then, by hypothesis, $G \in U^{k+1} \backslash T^k$; that is, by Lemma 3, G

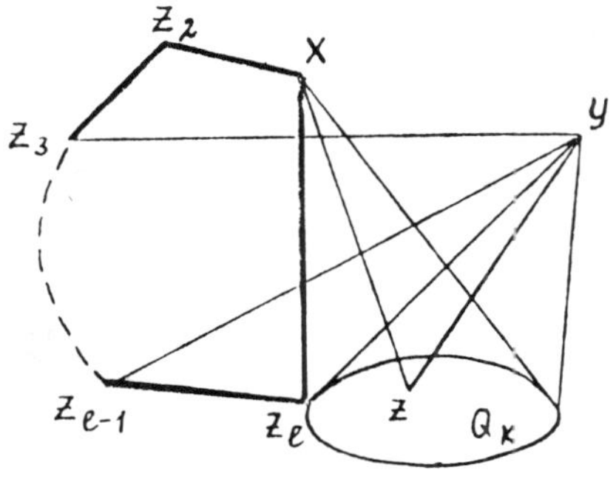

FIGURE 2

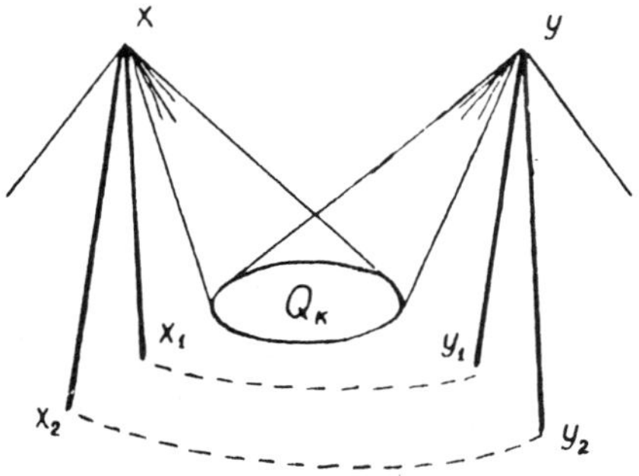

FIGURE 3

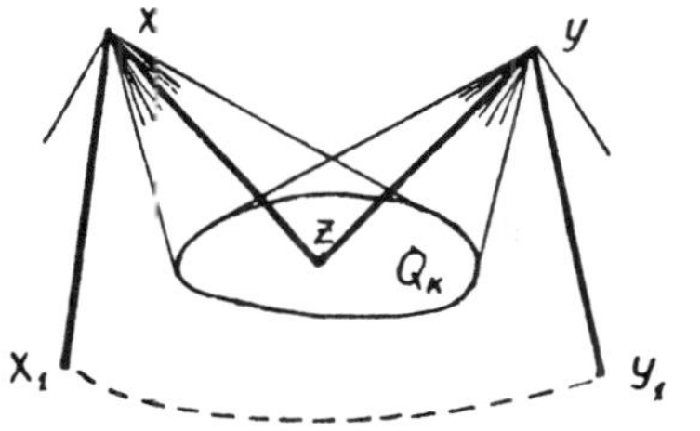

FIGURE 4

$k+1$ colors occurs on the vertices of C_l at least once, since all k-color subgraphs are $(k-1)$-trees. Consequently, $l \geq k+1$.

If the cycle C_l does not pass through the vertices x and y, then this cycle does not have a chord in G'. This contradicts the fact that G' is a k-tree. Thus, C_l passes through at least one of the vertices x and y. Therefore, C_l passes through no more than one vertex of Q_k. We have three cases.

Case 1. Suppose that C_l passes through x, but not y. We denote the vertices of the cycle in succession by $z_1 = x,\ z_2, \ldots, z_l$ (see Figure 2). Since z_2 and z_l are not adjacent, one of them, say z_2, does not belong to Q_k. The vertex y, which has the same color as x, in view of Lemma 2, is not adjacent to z_2, but it is adjacent to all the vertices $z_3, \ldots, z_{l-1}$; otherwise in the k-tree G' there would be a cycle without chords and of length ≥ 4.

Consider a vertex z of the k-clique Q_k that has the same color as z_2. Then the vertices $x,\ z,\ y,\ z_3$ and z_2 generate in G a 3-color cycle of length 5. Depending on whether there is a chord (z, z_3) or not, we have a cycle C_4 or C_5 without chords, in which only 3 colors occur, which is impossible, since in any cycle without chords in G all $k+1 \geq 4$ colors must occur. This gives a contradiction.

Case 2. Suppose that the cycle C_1 passes through x and y, but not through a vertex of the clique Q_k. Let $x_1,\ x_2$ and $y_1,\ y_2$ denote the vertices of C_l joined to

x and y respectively (see Figure 3). By Lemma 2, we have $x_i \neq y_j$, $i, j = 1, 2$. Furthermore, we show that x_i and y_j are not adjacent, $i, j = 1, 2$. Without loss of generality we may assume that x_1 and y_1 are adjacent and have colors 2 and 3 respectively. Then in the 2-tree $G[V_1 \cup V_2 \cup V_3]$ the minimal separating set of the ends of the hyperedge $[xQ_2y]$ consists of no fewer than three vertices, despite Lemma 4. From this it follows that the length of the given cycle C_l is not less than 8. And since x and y are not adjacent to any vertices of the cycle except x_i and y_j, $i, j = 1, 2$, there is also a cycle without chords in the k-tree G'. This gives a contradiction.

Case 3. Suppose that the cycle C_l passes through x and y and one of the vertices (call it z) of Q_k. As in Case 2, the vertices x_1 and y_1 respectively adjacent to x and y (distinct from z) are not adjacent; that is, the cycle C_l has length $l \geq 6$ (see Figure 4). Since C_l does not have chords, the vertices x and y are not adjacent to any vertices of C_l between x_1 and y_1. Therefore, after the identification of x and y in the k-tree G' there appears a cycle without chords and of length ≥ 4. This gives a contradiction.

Thus, our assumption that the graph $G \in U^{k+1}$ has a cycle without chords is false. By Lemma 3, this implies that G is a k-tree; that is, we obtain a contradiction with our original assumption. This proves Theorem 3.

COROLLARY 3.1. *For $k \geq 2$ a $(k+1)$-colorable graph G is a k-tree if and only if for some $(k+1)$-coloring of it every $(m+1)$-color subgraph is an m-tree, where $2 \leq m \leq k$.*

REMARK. Generally speaking, Theorem 3 is false for $k = 1$ or 2. For $k = 1$ a counterexample is any two-complete graph with at least one cycle. For $k = 2$ we construct counterexamples starting from an example given in [7].

Let G be a graph with $p \geq 6$ vertices, obtained from the complement of the cycle C_6 by adding $p - 6$ new vertices of degree 2 that are adjacent to the ends of certain edges. Obviously, G can be colored in three colors in two essentially different ways. For either of them all the 2-color subgraphs are trees, in other words, 1-trees. But G is not a 2-tree. Here it is relevant to mention a more general result of [8]. Nieminen gave examples for any $n \geq 2$ of nonuniquely n-colorable graphs in which the $(n-1)$-color subgraphs for any n-coloring are uniquely $(n-1)$-colorable.

For the proof of the main theorem we also need the following properties of chromatic polynomials.

Let G be an n-chromatic graph with p vertices, q edges and t triangles, and let $f(G, \lambda)$ be its chromatic polynomial.

PROPERTY 1 (SEE [9]). *The coefficient of λ^{p-1} in $f(G, \lambda)$ is $-q$.*

PROPERTY 2 (SEE [10] OR [11]). *The coefficient of λ^{p-2} in $f(G, \lambda)$ is $\binom{q}{2} - t$.*

PROPERTY 3 (SEE [9]). *The graph G can be colored in n colors in exactly $f(G, n)/n!$ ways, up to permutations of the colors.*

PROOF OF THEOREM 1. *Necessity* follows from the definition of a k-tree. In fact, a k-tree G with p vertices is obtained from a k-tree G' with $p-1$ vertices by adding a new vertex x adjacent to every vertex of some k-clique Q_k of G'. Then the vertex x of the graph G can take any color other than those of the vertices of Q_k, so x can be colored in $\lambda - k$ ways. Thus, $f(G,\lambda) = (\lambda - k) \cdot f(G', \lambda)$. Then by induction on the number of vertices, using the relation $f(K_{k+1}, \lambda) = \lambda(\lambda - 1) \cdots (\lambda - k)$, we obtain

$$f(G,\lambda) = \lambda(\lambda-1)\cdots(\lambda-k+1)(\lambda-k)^{p-k}.$$

Sufficiency. Suppose that G is a graph for which

$$f(G,\lambda) = \lambda(\lambda-1)\cdots(\lambda-k+1)(\lambda-k)^{p-k}.$$

For $k = 0$ the assertion is obvious. A proof for the case $k = 1$ can be found in [2]. Suppose that $k \geq 2$. Since the coefficient of λ^{p-1} in $f(G,\lambda)$ is $-pk + \binom{k+1}{2}$, by Property 1, $q = pk - \binom{k+1}{2}$. Moreover, $f(G, k+1)/(k+1)! = 1$, so that, by Property 3, G is uniquely $(k+1)$-colorable; that is, $G \in U^{k+1}$. Calculating the coefficient of λ^{p-2} in $f(G,\lambda)$ and using Property 2, we obtain

$$t(G) = p \cdot \binom{k}{2} - 2\binom{k+1}{3}. \tag{*}$$

Then for $k = 2$ it follows immediately from Theorem 2 that G is a 2-tree. We show that for $k \geq 3$ every 3-chromatic subgraph of G is a 2-tree. Suppose that this is not so. By Corollary 2.1 we have

$$t(G) < \sum_{\alpha,\beta,\gamma} (p_\alpha + p_\beta + p_\gamma - 2),$$

where $p_1 + \cdots + p_{k+1} = p$. Here this summation is carried out over all unordered triples of distinct elements of the set $\{1, 2, \ldots, k+1\}$. It is not difficult to check that

$$\sum (p_\alpha + p_\beta + p_\gamma - 2) = p \cdot \binom{k}{2} - 2\binom{k+1}{3}.$$

This contradicts equality $(*)$. Now using Corollary 3.1, we conclude that G is a k-tree. This proves Theorem 1.

The author would like to express his thanks to D. S. Mel'nikov and O. V. Borodin for their interest and advice.

BIBLIOGRAPHY

1. A. A. Zykov, *Graph theory*, "Nauka", Novosibirsk, 1969. (Russian)
2. Frank Harary, *Graph theory*, Addison-Wesley, 1969.
3. Frank Harary and Edgar M. Palmer, *On acyclic simplicial complexes*, Mathematika **15** (1968), 115–122.
4. L. W. Beineke and R. E. Pippert, *Properties and characteristics of k-trees*, Mathematika **18** (1971), 141–151.
5. Donald J. Rose, *On simple characterizations of k-trees*, Discrete Math. **7** (1974), 317–322.

6. G. A. Dirac, *On rigid circuit graphs*, Abh. Math. Sem. Univ. Hamburg **25** (1961), 71–76.

7. Frank Harary, S. T. Hedetniemi, and R. W. Robinson, *Uniquely colorable graphs*, J. Combinatorial Theory **6** (1969), 264–270.

8. Juhani Nieminen, *A note on uniquely colorable graphs*, Glasnik Mat. Ser. III **9(29)** (1974), 193–195.

9. Ronald C. Read, *An introduction to chromatic polynomials*, J. Combinatorial Theory **4** (1968), 52–71.

10. G. H. J. Meredith, *Coefficients of chromatic polynomials*, J. Combinatorial Theory Ser. B **13** (1972), 14–17.

11. E. J. Farrell, *On chromatic coefficients*, Discrete Math. **29** (1980), 257–264.

Translated by E. J. F. PRIMROSE

Amer. Math. Soc. Transl.
(2) Vol. **132**, 1986

Directed Cycles of Any Length in Digraphs with Large Semidegrees

UDC 519.1

S. KH. DARBINYAN

We consider finite directed graphs (digraphs) without loops or parallel arcs. All concepts and notations not defined here can be found in [1]. A digraph with p vertices containing a directed cycle of any length $k \in [3, p]$ is called *pancyclic*. A digraph is called *k-biregular* if $\text{od}(x) = \text{id}(x) = k$ for each of its vertices x. In [2] it is proved that if in a p-vertex strongly connected digraph G the sum of the degrees of any two nonadjacent vertices is not less than $2p-1$ then G is pancyclic, except for certain simple cases. In [3] the pancyclicity of the $(2n+1)$-vertex $(n-1)$-biregular directed graphs for $n \geq 8$ is proved. More detailed information on pancyclic digraphs can be found in the survey article [4]. In [5] Thomassen studied the Hamiltonicity of p-vertex digraphs with minimum degree $p-1$. In particular a result was obtained which partially describes the structure of strongly-connected non-Hamiltonian digraphs of this type, and it was proved that if in a $(2n+1)$-vertex digraph G the minimum out-degree and in-degree is not less than n then G (except for certain simple cases) is Hamiltonian. In the present note their pancyclicity is studied.

Let G be a digraph. We denote by $V(G)$ the set of vertices of G and by $E(G)$ the set of its arcs. For $A, B \subseteq V(G)$ and $x \in V(G)$ we introduce the notation

$$E(A \to B) = \{yz \in E(G) | y \in A, z \in B\},$$
$$O(x) = \{y \in V(G) | xy \in E(G)\},$$
$$I(x) = \{y \in V(G) | yx \in E(G)\},$$
$$E(A, B) = E(A \to B) \cup E(B \to A),$$
$$d(x, A) = |E(\{x\}, A)|.$$

1980 *Mathematics Subject Classification* (1985 *Revision*). Primary 05C20, 05C38; Secondary 05C45.

Translation of Akad. Nauk Armyan. SSR Dok. **75** (1982), 147–152; MR **84g**:05066.

The number $d(x) = \mathrm{id}(x) + \mathrm{od}(x)$ is called the *degree* of the vertex x. The notation $A \to B$ means that if $y \in A$ and $z \in B$ then $yz \in E(G)$. If $C \subseteq V(G)$ and $A \to B$ and $B \to C$, we will write $A \to B \to C$. A directed cycle of length k will be denoted by $\vec{C}_k$. If G_1 is an ordinary graph, then G_1^* denotes the symmetric digraph obtained from G_1.

The following generally-known simple properties of digraphs will often be used in the course of the proofs of Theorems 1 and 2.

LEMMA 1. *Let C be a directed cycle of length m in the digraph G and $x \in V(G) \setminus V(C)$. If $d(x, V(C)) \geq m+1$ then G contains a directed cycle of any length $r \in [2, m+1]$.*

LEMMA 2. *Let P: $x_1x_2\cdots x_m$, $m \geq 2$, be a path in the digraph G, and let $x \in V(G) \setminus V(P)$. Let one of the following conditions hold:*

a) $d(x(V(P)) \geq m+2$.

b) $d(x, V(P)) \geq m+1$ and $xx_1 \notin E(G)$ or $x_mx \notin E(G)$.

c) $d(x, V(P)) \geq m$, $xx_1 \notin E(G)$ and $x_mx \notin E(G)$.

Then there exists $i \in [1, m-1]$ such that x_ixx_{i+1} (in this case one says that the path $x_1x_2\cdots x_ixx_{i+1}\cdots x_m$ is obtained from P by extension by means of the vertex x).*

THEOREM 1. *Let G be a p-vertex strongly connected non-Hamiltonian digraph such that the sum of the degrees of any two of its nonadjacent vertices is not less than $2p-2$. If $\vec{C}_m = x_1x_2\cdots x_mx_1$ is the longest directed cycle in G, then the following assertions are true:*

I. Any two vertices from $V(G) \setminus V(\vec{C}_m)$ are adjacent, any vertex $x \notin V(\vec{C}_m)$ has degree not greater than $p-1$, and every strong component G_i, $1 \leq i \leq s$, of the subgraph $\langle V(G) \setminus V(\vec{C}_m)\rangle$ is a complete digraph. Furthermore, if the digraph G is 2-connected, then $\langle V(G) \setminus V(\vec{C}_m)\rangle$ is a transitive tournament. (An analogous result for digraphs with minimum degree not less than $p-1$ is obtained in [5].)

II. G contains a directed cycle of any length $k \in [2, m]$ unless $G = K^*_{[p/2],[p/2]+1}$.

We give only the outline of a proof of Theorem 1.

Let $G_0 = \langle V(\vec{C}_m)\rangle$ and $P = \langle V(G) \setminus V(\vec{C}_m)\rangle$, and suppose that $G_1, \ldots, G_s$ are numbered so that for all i and j, $1 \leq i < j \leq s$, we have

$$E(V(G_j) \to V(G_i)) = \varnothing.$$

Consequently, by the strong connectedness of G we have

$$E(V(G_0) \to V(G_1)) \neq \varnothing, \; E(V(G_s) \to V(G_0)) \neq \varnothing. \tag{1}$$

From this it follows that one can choose vertices $u \in V(G_{k_1})$, $v \in V(G_{k_2})$, and $x_\sigma, x_\tau \in V(G_0)$, where $x_\sigma u$, $vx_\tau \in E(G)$, $1 \leq k_1 \leq k_2 \leq s$, such that the path $x_\sigma x_{\sigma+1}\cdots x_{\tau-1}x_\tau$ has the smallest possible length $\mu+1$ and $x_\sigma \neq x_\tau$ (for $x_\sigma = x_\tau$ the truth of the theorem is evident). We extend the path $x_\tau x_{\tau+1}\cdots x_\sigma$

* *Translator's note.* x_ixx_{i+1} means that $x_ix \in E(G)$ and $xx_{i+1} \in E(G)$.

by means of the vertices $x_{\sigma+1}, \ldots, x_{\sigma+\mu}$ as much as possible. After this there will still be certain vertices $y_1, \ldots, y_d \in \{x_{\sigma+1}, \ldots, x_{\sigma+\mu}\}$, $1 \le d \le \mu$, which do not belong to the extended path. Further, with the help of the minimality of μ and Lemma 2 the following assertion can be proved:

1°. *G_{k_1} and $G'_{k_1} = \langle\{x_{\sigma+1}, \ldots, x_{\sigma+\mu}\}\rangle$ are complete digraphs; $\mu = d$, $k_1 = k_2$; for any vertices $y \in \{y_1, \ldots, y_d\}$ and $z \in V(G_{k_1})$ we have $d(z) \le p-1$, $d(z, V(G_0)) = m - \mu + 1$ and $d(y, V(G_0)) = m + \mu - 1$; and, finally,*

$$\bigcup_{i=1}^{k_1-1} V(G_i) \to V(G_{k_1}) \cup V(G'_{k_1}) \to \bigcup_{i=k_1+1}^{s} V(G_i).$$

Therefore, for the proof of the first part of Theorem 1 it suffices to consider the case where $s \ge 2$ and $k_1 \ge 2$. From (1) and 1° it follows that we can choose vertices $u_1 \in v(G_{k_3})$, $v_1 \in V(G_{k_4})$ and $x_{\sigma_1}, x_{\tau_1} \in V(G_0)$, where $1 \le k_3 \le k_1 \le k_4 \le s$ and $x_{\sigma_1}u_1, v_1x_{\tau_1} \in E(G)$, such that the path $x_{\sigma_1}x_{\sigma_1+1}\cdots x_{\tau_1}$ have the smallest possible length $\mu_1 + 1$. If $G'_{k_3} = \langle\{x_{\sigma_1+1}, x_{\sigma_1+2}, \ldots, x_{\sigma_1+\mu_1}\}\rangle$, the proof of the assertion is analogous to that of 1° for the subgraphs G_{k_3} and G'_{k_3}. Further, the analogous assertion for every G_i and G'_i, $1 \le i \le s$, is proved in the same way. We now have enough of the structure of the digraph G to be able to prove the first part of Theorem 1.

Further, if $|V(G_i)| \ge 2$ or $|V(G'_i)| \ge 2$, then clearly $C_r \subset G$ for all $r \in [2, m]$. Suppose that $|V(G_i)| = 1$ for all $i \in [1, s]$. Then by the first part of Theorem 1 the subgraph Q is a transitive tournament. Let $\{x\} = V(G_1)$, and let $y_1, \ldots, y_k$ be those vertices of the directed cycle $\vec{C}_m$ which are not adjacent to the vertex x, numbered in such a way that the vertex y_i immediately precedes y_{i+1} in the direction of the directed cycle (the indices of the vertices y_i are taken modulo k)*. For any vertex y_i, $1 \le i \le k$, we denote by A_{2i} (by A_{2i+1}) the set of those vertices of the contour $\vec{C}_m$ which immediately precede (follow) y_i and $A_{2i} \to x$, $E(x \to A_{2i}) = \varnothing$ (respectively, $x \to A_{2i+1}$, $E(A_{2i+1} \to x) = \varnothing$) and A_j, $1 \le j \le 2k$, are maximal (the indices of the sets A_j are taken modulo $2k$).**

Suppose that $C_r \not\subset G$ for some $r \in [2, m-1]$. Then it is not hard to show that for any $i \in [1, m]$

$$|E(x \to x_i)| + |E(x_{i+r-2} \to x)| = 1. \tag{2}$$

Let $a_j = |A_j|$ and suppose that $a_1 = \max_{1 \le j \le 2k}\{a_j\}$. For any $l \in [1, k]$, consider the following intervals:

$$I_l = \left[\sum_{i=2}^{2l-1} a_i + 2l, \sum_{i=1}^{2l} a_i + 2l\right].$$

* *Translator's note.* This means that y_1 follows y_k, and not that y_0 follows y_{k-1} as the strict number-theoretic meaning of modulo k would imply.

** *Translator's note.* In other words, among the vertices of $\vec{C}_m$ that lie between y_{i-1} and y_i, A_{2i-1} is the set of vertices z such that $xz \in E(G)$ and $zx \notin E(G)$, and A_{2i} is the set of vertices z such that $xz \notin E(G)$ and $zx \in E(G)$.

Using (2) and the maximality of a_1 we can show that $r \notin \bigcup_1^k I_l$ and that for any $q \in [1, k]$ we have

$$r = \sum_{i=2q-1}^{2q+2l-2} a_i + 2l + 1.$$

Thence, with the help of the already-obtained structure of G it can be shown that if $E(\langle\{x, y_1, \dots, y_k\}\rangle) \neq \varnothing$ then $\vec{C}_r \subset G$, and if $E(\langle\{x, y_1, \dots, y_k\}\rangle) = \varnothing$ then $G = K^*_{[p/2],[p/2]+1}$.

THEOREM 2. *Suppose that G is a $(2n+1)$-vertex digraph with minimum out-degree and in-degree $\geq n$. Then either G is pancyclic, or $G \in \{C_5^*, D_5, d_7, [(K_n \cup K_n) + K_1]\}^*$, or*

$$K^*_{n,n+1} \subseteq G \subseteq [K_n + \overline{K}_{n+1}]^*$$

(the digraphs D_5 and D_7 are shown in the diagram).

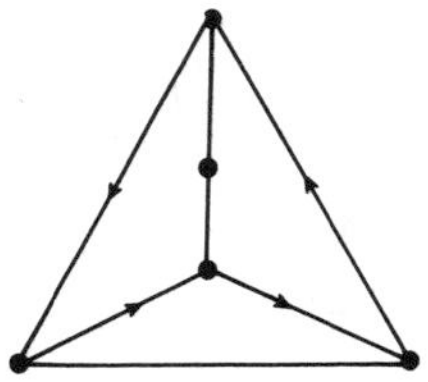

D_5

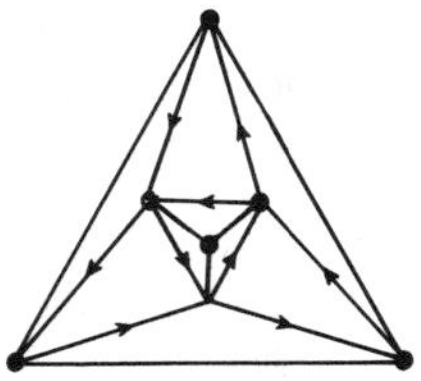

D_7

Thomassen showed in [5] that under the conditions of Theorem 2 the digraph G is Hamiltonian.

Sketch of proof. One first proves that G contains a directed cycle of any length $r \in [3, 5]$. After that one proves that $\vec{C}_{2n} \subset G$. To do this, one considers a maximal non-Hamiltonian directed cycle $\vec{C}_m$ of the digraph G and supposes that $m \leq 2n - 1$. With the help of the arguments used in the proof of Lemma 3 in [6] it can be shown that $V(G) \setminus V(\vec{C}_m) = \{u, v\}$, uvu and $d(u) = d(v) = 2n$. After that, by considering the structure of G one shows that the vertex u (v) is not adjacent to two neighboring vertices of the directed cycle $\vec{C}_{2n}$ at the same time. Consequently, there exists $j \in [1, m]$ such that $E(u, \{x_j, x_{j+1}\}) = \varnothing$ and

$$O(u) = I(u) = \{v, x_{j+2}, x_{j+4}, \dots, x_{j-1}\}.$$

From this we obtain $|E(v, x_j)| = 2$ and $\vec{C}_{m+1}\colon x_j v u x_{j+2} \cdots x_{j-1} x_j$, which contradicts the assumed maximality of m.

Let $\vec{C}_{2n}$ be $x_1 x_2 \cdots x_{2n} x_1$, and suppose that $x \notin V(C_{2n})$ and $C_m \not\subset G$. Then clearly $d(x) = 2n$, and for any $i \in [1, 2n]$ we have

$$|E(x \to x_i)| + |E(x_{i+m-2} \to x)| = 1. \tag{3}$$

We consider the following possible cases.

Case 1. Any vertex of the directed cycle $\vec{C}_{2n}$ is adjacent to the vertex x.

We denote by $M_1, \dots, M_k$ (by $N_1, \dots, N_k$) the set of vertices of the directed cycle $\vec{C}_{2n}$ such that $x \to M_i$ $(N_i \to x)$, the vertices of the set M_i (N_i) constitute a path in the direction of the directed cycle $\vec{C}_{2n}$, and the sets are maximal. Furthermore M_i immediately precedes N_i in the directed cycle $\vec{C}_{2n}$ (everywhere the indices of these sets are taken modulo k).*

Let $n_j = |N_j|$, $m_j = |M_j|$, $1 \le j \le k$, and

$$m_1 = \max_{1\le j\le k}\{m\} \ge \max_{1\le j\le k}\{n_j\}.$$

For any $l \in [1, k]$ we consider the following intervals:

$$I_l = \left[\sum_{j=1}^{l-1} n_j + \sum_{j=2}^{l} m_j + 3,\ \sum_{j=1}^{l}(m_j + n_j) + 1\right].$$

It is not difficult to see that $m \notin \bigcup_1^k I_l$ and that for some $s \in [2, k]$

$$m = \sum_{j=i}^{i+s-2}(n_j + m_j) + 2.$$

After this one considers the separate cases $m_1 \ge 2$ and $m_1 = 1$ and shows that $\vec{C}_m \subset G$, which contradicts the assumptions.

Case 2. There exists a vertex x_i of the directed cycle $\vec{C}_{2n}$ which is not adjacent to the vertex x.

Without loss of generality it can be assumed that $i = 2n = p$ and $x_{p-k}xx_a$, where $k \ge 1$, $a \ge 1$, $E(x, x_1) \ne \varnothing$, and

$$E(\{x_{p-k+1}, x_{p-k+2}, \dots, x_p\} \to x) = E(x \to \{x_1, x_2, \dots, x_{a-1}\}) = \varnothing.$$

From this and (3) it follows that $m - 2 \le p - k - a + 1$.

Further, by the supposition that $\vec{C}_m \not\subset G$ and Lemma 2 it can be shown that

2°. *For any s and t, where $p - k - 1 \le s < t \le p$, the condition $x_s x_t \in E(G)$ holds if and only if $t = s + 1$.*

From this assertion it follows, in particular, that

$$E(\{x_p\} \to \{x_2, x_3, \dots, x_{a+1}\}) = E(\{x_{p-k-1}, x_{p-k}, \dots, x_{p-2}\} \to \{x\}) = \varnothing.$$

Let P be $x_{j_1}x_{j_1+1}\cdots x_{p-k-1}x_{p-k}$, where $j_1 = a$. We suppose that there exists a path $P_{i-1}\colon x_{j_{i-1}}x_{j_{i-1}+1}\cdots x_{j_{i-1}+m-3}$, where $i \ge 2$. If $j_{i-1} + m - 3 \le p-k-m+2$ and there exists $j_i \in [j_{i-1}+1, j_{i-1}+m-3]$ such that $xx_{j_i} \in E(G)$ and $xx_{j_i-1} \notin E(G)$, then consider the path $P_i\colon x_{j_i}x_{j_i+1}\cdots x_{j_i+m-3}$ (the numbers j_i are taken as large as possible). Let $r-1$ be the maximum number of such paths P_i and let P_r be $x_{p-k-m+3}x_{p-k-m+4}\cdots x_{p-k}$.

If $j_{r-1}+m-3 \ge p-k-m+3$, we say that the path P is covered by the paths $P_1, \dots, P_r$, and if $j_{r-1} + m - 3 < p - k - m + 3$, then the path P is not covered by the paths $P_1, \dots, P_r$. We note that any path P_i, $1 \le i \le r$, has length $m - 3$.

We consider the following subcases of case 2.

---

* *Translator's note*: The vertices of $\vec{C}_{2n}$ are divided into intervals $M_1, N_1, M_2, N_2, \dots, M_k, N_k$ of vertices alternately adjacent *from* x and adjacent *to* x.

Subcase 2.1. The path P is covered by $P_1, \ldots, P_r$. In this case, with the help of Lemma 2 it can be shown that $x_{j_1}x_p x_{j-k}$, $j_1 = k = 1$, and

$$E(\{x_p\}, \{x_{j_1+1}, x_{j_1+3}, \ldots, x_{i_r+m-4}\}) = \varnothing,$$
$$\{x_p\} \to \{x_{j_1+2}, x_{j_1+4}, \ldots, x_{j_r+m-5}\} \to \{x_p\}.$$

Hence m is odd. This means that Subcase 2.1 holds for the directed cycle $xx_1x_2\cdots x_{p-1}x$. Therefore

$$E(\{x\}, \{x_2, x_4, \ldots, x_p\}) = \varnothing,$$
$$\{x\} \to \{x_1, x_3, \ldots, x_{p-1}\} \to \{x\}.$$

Further, in the same way we obtain $E(\langle\{x, x_2, x_4, \ldots, x_p\}\rangle) = \varnothing$; that is,

$$K^*_{n,n+1} \subseteq G \subseteq (K_n + \overline{K}_{n+1})^*.$$

Subcase 2.2. The path P is covered by $P_1, \ldots, P_r$. In this case, again with the help of the special construction of the digraph G, the existence of a directed cycle of length m can be shown, which contradicts the assumption that $\vec{C}_m \not\subset G$.

REFERENCES

1. Frank Harary, *Graph theory*, Addison-Wesley, 1969.
2. S. Kh. Darbinyan, *Panconnected and pancyclic digraphs*, author's summary of dissertation, Minsk, 1981. (Russian)
3. S. Kh. Darbinyan anrd K. M. Mosesyan, *On the pancyclicity of regular digraphs*, Akad. Nauk Armyan. SSR Dokl. **67** (1978), 208–211. (Russian)
4. J.-C. Bermond and C. Thomassen, *Cycles in digraphs—a survey*, J. Graph Theory **5** (1981), 1–43.
5. Carsten Thomassen, *Long cycles in digraphs*, Proc. London Math. Soc. (3) **42** (1981), 231–251.
6. S. Kh. Darbinyan, *On pancyclic digraphs*, Preprint, Computing Center, Acad. Sci. Armenian SSR and Erevan State Univ., Erevan, 1979. (Russian) R. Zh. Mat. **1980**, 2B644.

Translated by T. R. WALSH

Amer. Math. Soc. Transl.
(2) Vol. **132**, 1986

On the Minimum of the Hadwiger Number for Graphs with a Given Mean Degree of Vertices

UDC 519.1

A. V. KOSTOCHKA

Introduction

The Hadwiger number $\eta(G)$, as one of the interesting topological characteristics of graphs [1], has attracted particular attention in connection with Hadwiger's conjecture that for any graph G its Hadwiger number is not less than its chromatic number $\chi(G)$. A special case of this conjecture is the four-color problem.

Let us recall the concept of the Hadwiger number.

An elementary contraction of a graph G is one of the following operations:

a) removal of an edge;

b) removal of an isolated vertex;

c) replacement of two adjacent vertices v_1 and v_2 by a new vertex v_3 that is joined by an edge to every vertex adjacent in G to at least one of the vertices v_1 and v_2.

We say that a graph G is contracted to a graph H if H can be obtained from G by a sequence of elementary contractions. The Hadwiger number $\eta(G)$ of a graph G is the number of vertices of the maximal complete graph to which G is contracted.

A number of authors (see [5]–[7], for example) have been interested in how small a Hadwiger number a graph with a given mean degree of vertices can have. Mader [8] showed that for graphs with mean degree of vertices $2k$ the minimum Hadwiger number is greater than $k/8\log_2 k$. The aim of the present paper is to show that this minimum is a quantity of order $k/\sqrt{\ln k}$.

Let us state the problem more precisely.

1980 *Mathematics Subject Classification* (1985 *Revision*). Primary 05C35; Secondary 05C15.

Translation of Metody Diskret. Analiz. Vyp. **38** (1982), 37–58; MR **85c**:05018.

Let $\mathcal{D}_{n,k}$ (respectively, $\mathcal{E}_{n,k}$) be a set of graphs with n vertices ($n \geq k$) and with at least kn (respectively, more than $kn - \binom{k+1}{2}$) edges; and let $\mathcal{D}_k$ ($\mathcal{E}_k$) $= \bigcup_{k=n}^{\infty} \mathcal{D}_n$ ($\mathcal{E}_n$).

Our aim is to investigate $\eta(k) \doteq \min_{G \in \mathcal{D}_k} \eta(G)$ (the symbol $\doteq$ means "equals by definition").

Earlier (see [5], [6], or [8], for example), it was primarily $\eta_1(k) \doteq \min_{G \in \mathcal{E}_k} \eta(G)$ that was investigated. The result of Mader [8] mentioned above can be stated as follows: $\eta_1(k) > k/8 \log_2 k$. However, the quantities $\eta(k)$ and $\eta_1(k)$ do not differ in order. Obviously, $\eta(k) \geq \eta_1(k)$.

Let us put $w(k) = \min_{\{G/\chi(G) \geq k\}} \eta(G)$; then Hadwiger's conjecture is equivalent to the assertion that $w(k) \geq k$ for all natural numbers k. Wagner [3] showed that $w(k) \geq 4 + \log_2((k-1)/3)$. From a result of Mader [8] it follows that $w(k) > k/16 \log_2 k$.

The main results of this paper are the following.

ASSERTION 1. *For any sufficiently large* k

$$\eta_1(k) \leq 2.25k/\sqrt{\ln k}; \tag{a}$$

$$\eta(k) \leq 3.15k/\sqrt{\ln k}. \tag{b}$$

THEOREM. *For any* $k \geq 2$

$$\eta(k) \geq \eta_1(k) \geq 0.064k/\sqrt{\ln k} + 1.$$

COROLLARY 1. *For any* $k \geq 3$

$$w(k) \geq 0.032k/\sqrt{\ln k}.$$

PROOF. Every k-chromatic graph G contains a color-critical subgraph G' for which the degree of every vertex is at least $k-1$. Therefore, $G' \in \mathcal{D}_{[(k-1)/2]}$. (Here and later $[x]$ denotes the integral part of the real number x.) By the theorem,

$$\eta(G) \geq \eta(G') \geq 0.064(k/2 - 1)/\sqrt{\ln(k/2)} + 1 \geq 0.032k/\sqrt{\ln k}.$$

COROLLARY 2. *Hadwiger's conjecture is true for almost all graphs with* n *vertices.*

PROOF. It is known [9] that, for almost all graphs G with n vertices, $\chi(G) \leq n/\ln n$. It is also known that, for almost all graphs G with n vertices, $|E(G)| \geq n^2/6$. Hence, by the theorem, $\eta(G) \geq 0.064n/6\sqrt{\ln n}$. For large n we have $0.064n/6\sqrt{\ln n} > n/\ln n$.

REMARK. The result of Corollary 2 was obtained independently by Bollobás, Catlin, and Erdős.

COROLLARY 3. *Hadwiger's conjecture is true for almost all graphs with* n *vertices and* kn *edges if* n *is sufficiently large.*

PROOF. It is known ([4], Chapter 16, Exercise 9) that, if k is large, for almost all graphs G with n vertices and kn edges, $\chi(G) \leq 3k/\log_2 k$. By the theorem, for any graph G of this class, $\eta(G) \geq 0.064k/\sqrt{\ln k}$. Therefore, for large k we have $\eta(G) \geq \chi(G)$.

COROLLARY 4.

$$\min_{|V(G)|=n} (\eta(G) + \eta(\overline{G})) = O(n/\sqrt{\ln n}),$$

where $V(G)$ is the set of vertices of G, and $\overline{G}$ is the complement of G.

PROOF. Suppose that $n \geq 2$ and $|V(G)| = n$. Then the number of edges of either G or $\overline{G}$ is greater than $(n/6)n$. Therefore, by the theorem, $\max\{\eta(G), \eta(\overline{G})\} \geq 10^{-2}n/\sqrt{\ln n}$. On the other hand, putting $p = 0.5$ in the proof of Assertion 2 (in §1, below), we see that, for almost all graphs G with n vertices, $\eta(G) \leq n/\sqrt{\ln n}$. Hence there is a graph G_0 such that $\eta(G) = n/\sqrt{\ln n}$ is true for both G_0 and $\overline{G}_0$. Consequently, $\eta(G_0) + \eta(\overline{G}_0) \leq 2n/\sqrt{\ln n}$.

COROLLARY 5. *Let $\nu(k)$ be the smallest Hadwiger number that a k-connected graph can have. Then $\nu(k) = O(k/\sqrt{\ln k})$.*

Corollary 5 follows from the obvious inequality $2\nu(k) \geq \eta(k)$ and a result of Mader [8] that $\nu(k) \leq 2\eta(k)$.

Let us agree on the following notation. For any graph G let $V(G)$, $E(G)$ and $\overline{G}$ denote respectively the set of vertices, the set of edges and the complement of G. We also put $n(G) \doteq |V(G)|$ and $m(G) \doteq |E(G)|$.

If $V_0 \subset V(G)$, let $G(V_0)$ and $G\backslash V_0$ denote the subgraphs of G generated by the sets of vertices V_0 and $V(G)\backslash V_0$, respectively. If $v \in V(G)$, then $N_G(v) \doteq \{w \in V(G) | (v, w) \in E(G)\}$, $s_G(v) \doteq |N_G(v)|$, and $\sigma(G) \doteq \max_{v \in V(G)} s_G(V)$.

§1. Upper bounds for $\eta(k)$ and $\eta_1(k)$

A *pseudocoloring* of a graph G with r colors is defined as any partition of $V(G)$ into subsets $V_1, \ldots, V_r$ such that for any $1 \leq i < j \leq r$ there is an $e \in E(G)$ joining V_i and V_j. The *pseudochromatic number* $\tau(G)$ of a graph G is defined as the largest r for which there is a pseudocoloring of G with r colors.

A contraction of a graph G onto a complete graph K_r can be treated as a map $\varphi: V_0 \overset{\text{onto}}{\to} V(K_r)$, where $V_0 \subset V(G)$. If for each $v_i \in V(K_r)$ the vertices of the set $\varphi^{-1}(v_i)$ are colored with the ith color, we obtain a pseudocoloring of G with r colors. Consequently, $\tau(G) \geq \eta(G)$ for any graph G. Therefore, Assertion 1 follows from the next result.

ASSERTION 2. *If k is sufficiently large, then*
(a) $\min_{G \in \mathcal{E}(k)} \tau(G) < 2.25k/\sqrt{\ln k}$;
(b) $\min_{G \in \mathcal{D}(k)} \tau(G) < 3.15k/\sqrt{\ln k}$.

PROOF. Suppose that $0.5 \leq p < 1$ and that $G_{n,p}$ is a random variable whose admissible values are elements of the set of all n-vertex graphs with numbered vertices, and $P(e \in G_{n,p}) = p$ for every edge e and for different edges these probabilities are independent.

Let us compute the probability $\Psi(k_1, \dots, k_r)$ of the appearance of n-vertex graphs for which a fixed partition $\mathcal{M}$ of the set $V(G)$ into r subsets $M_1, \dots, M_r$ with $|M_i| = k_i$ $(i = 1, \dots, r)$ is a pseudocoloring. The probability that there is an edge between M_i and M_j is $1 - (1-p)^{k_i k_j}$. Hence,

$$\Psi(k_1, \dots, k_r) = \prod_{i=1}^{r-1} \prod_{j=i+1}^{r} (1 - (1-p)^{k_i k_j}).$$

By any method for finding a numerical extremum it is easy to establish that

$$\max_{\substack{k_1 + \dots + k_r = n \\ k_1, \dots, k_r \geq 1}} \Psi(k_1, k_2, \dots, k_r) = \Psi\left(\frac{n}{r}, \frac{n}{r}, \dots, \frac{n}{r}\right) = (1 - (1-p)^{n^2/r^2})^{r(r-1)/2}.$$

Since the number of partitions of an n-element set into r subsets is less than n^r, the probability of the appearance of an n-vertex graph G with $\tau(G) \geq r$ does not exceed

$$n^n (1 - (1-p)^{n^2/r^2})^{r(r-1)/2} < \exp\{n \cdot \ln n - (1-p)^{n^2/r^2} r^2/3\}. \qquad (*)$$

If for some $\delta > 1/(1-p)$ we have $r \geq n/\sqrt{\log_\delta n}$ then the right-hand side of $(*)$ does not exceed

$$\exp\left\{n \ln n - \frac{n^2 (1-p)^{\log_\delta n}}{3 \log_\delta n}\right\} = \exp\left\{n \left(\ln n - \frac{n^{1+\log_\delta(1-p)}}{3 \log_\delta n}\right)\right\} \underset{n \to \infty}{\longrightarrow} 0.$$

Suppose that $p = 2/3$ and $r \geq n/\sqrt{\log_{3.01} n}$. Then for large n the probability that $\tau(G_{n,p}) \geq r$ is less than 10^{-3}. On the other hand, by the de Moivre-Laplace theorem, for large n the probability that $m(G_{n,p}) \geq n^2/3$ is greater than 0.01. Hence there is a graph G_n such that $n(G_n) = n$, $m(G_n) \geq n^2/3$, and $\tau(G_n) < n/\sqrt{\log_{3.01} n}$. If n is divisible by 3 and $k = n/3$, then $G_n \in \mathcal{D}_k$ and

$$\tau(G_n) \leq \frac{3k}{\sqrt{\log_{3.01} 3k}} < \frac{3k\sqrt{\ln 3.01}}{\sqrt{\ln k}} \leq 3.15 \cdot \frac{k}{\sqrt{\ln k}}.$$

This proves (b).

Similarly, suppose that $p = 8/9$ and $r \geq n/\sqrt{\log_{9.2} n}$. Then, as above, for sufficiently large n there is a graph H_n such that $n(H_n) = n$, $m(H_n) \geq \frac{4}{9}n^2$, and $\tau(H_n) < n/\sqrt{\log_{9.2} n}$. If n is divisible by 3 and $k = 2n/3$, then $H_n \in \mathcal{E}_k$ and

$$\tau(H_n) \leq \frac{3k/2}{\sqrt{\log_{9.2}(3k/2)}} \leq \frac{3k\sqrt{\ln 9.2}}{2\sqrt{\ln k}} \leq 2.25 \frac{k}{\sqrt{\ln k}}.$$

This proves the assertion.

§2. Idea of the proof of the theorem

The main part of the proof consists in studying contractions of graphs in which the number of edges is substantially greater than in their complements.

The main role is played by a function of real variables

$$f(n;m;\beta) \doteq \frac{n}{2\sqrt{\ln n}}\left(\sqrt{\ln\frac{n^2}{2m}}-\beta\right).$$

For any graph G we put $f(G;\beta) \doteq f(n(G);m(G);\beta)$. By means of $f(G;\beta)$ we shall give lower bounds for $\eta(\overline{G})$. In considering differences of the form $f(G;\beta)-f(H;\beta)$ we need the function

$$\varphi(c;z;\beta) \doteq (c-z)\cdot\sqrt{\ln\frac{c-z}{1+\ln c-\beta\sqrt{\ln c}-z}}-c\sqrt{\ln c}+\beta z.$$

In Lemmas 1–6 we determine analytic properties of the functions f and φ that we need in the proof of the later lemmas. Since the proofs of Lemmas 1–6 do not relate to graph theory, they are given in the Appendix.

The key result is Lemma 7. In it we prove that any n-vertex graph G with comparatively large minimal degree can be contracted to an $[n/2]$-vertex graph G' such that the number of edges in $\overline{G}'$ is substantially less than in $\overline{G}$. On the basis of Lemma 7 we show in Lemmas 8 and 9 that, for some β and restrictions on the number of edges in $\overline{G}$, the graph G can be contracted to a graph H with fewer edges such that $f(\overline{H};\beta) \geq f(\overline{G};\beta)$. Using this, in Lemma 10 we prove that, for graphs G whose complement contains less than 1/13 of the number of all edges,

$$\eta(G) \geq f(\overline{G};1.1).$$

Moreover, in Lemmas 11–15 we successively derive bounds analogous to those in Lemma 10 under weaker restrictions on the number of edges in $\overline{G}$. We use lower bounds for certain values of the function $\varphi(c;z;\beta)$, given in the table at the end of the paper.

The proof of the theorem really consists in reducing the general case to the case considered in Lemma 15.

LEMMA 1. *For any real $x \geq 2$ and $\beta \geq 3/5$*

$$x-1-2\ln x+2\beta\sqrt{\ln x} > 0.$$

LEMMA 2. *If $n \geq 4$, $2 \leq 2m \leq n$, and $0.6 \leq \beta < \sqrt{\ln n^2/2m}$, then $f(n-1;m;\beta)+1 \geq f(n;m;\beta)$, with f as defined at the beginning of* §2.

LEMMA 3. *Suppose that $2m \geq n \geq 4$ and*

$$0.6 \leq \beta < \sqrt{\ln\frac{n^2}{2m}}, \qquad \frac{n}{2} > \alpha \geq 1+\ln\frac{n^2}{2m}-\beta\sqrt{\ln\frac{n^2}{2m}}.$$

Then

$$f(n;m;\beta) \leq f(n-1;m-2m\alpha/n;\beta).$$

LEMMA 4. *Suppose that $c \geq 2$, $\beta \geq 0.6$, and $1 \leq z < 1+\ln c-\beta\sqrt{\ln c}$. Then $\varphi'_z(c;z;\beta) > 0$, where φ was defined at the beginning of* §2.

LEMMA 5. *Suppose $a \geq 2$, $\beta \geq 0.6$, $1 \leq z < 1 + \ln a - \beta\sqrt{\ln a}$, and $c \geq a$. Then $\varphi'_c(c; z; \beta) < 0$.*

LEMMA 6. *Suppose that $g(x) = e^{1.1x}/4 + 1.1x - x^2 - e^{1.1x - x^2}/4$. Then $g(x) > 0$ for any $x \geq 1.7$.*

§3. Lemma on contractions

LEMMA 7. *Suppose that $k \geq 6$ and that G is an n-vertex graph such that $\sigma(G) \leq n/k$. Then $\overline{G}$ can be contracted to an $[n/2]$-vertex graph $\overline{H}$ such that*

$$2m(H) \leq \frac{27k}{(k-2)^3(k-1)^2} n^2(H).$$

PROOF. We join each vertex $v \in V(G)$ by quasi-edges to those vertices not adjacent to v that are joined to v by more than $3n/(k-1)(k-2)$ paths of length 2. Then any vertex v is incident with no more than $n(k-2)/3k - 1$ quasi-edges. For otherwise the total number of paths of length 2 with v at one end would be greater than

$$\frac{3n}{(k-1)(k-2)} \cdot \frac{n(k-2) - 3k}{3k}.$$

However, the number of such paths does not exceed $\sigma(G)(\sigma(G) - 1) \leq n(n-k)/k^2$, and the inequality

$$\frac{3n}{(k-1)(k-2)} \cdot \frac{n(k-2) - 3k}{3k} > \frac{n(n-k)}{k^2}$$

is true for any $n \geq k \geq 6$.

Let G' denote the graph obtained from G by adding quasi-edges. Then

$$\sigma(G') \leq \frac{n}{k} + \frac{n(k-2)}{3k} - 1 = \frac{n(k+1)}{3k} - 1 < \frac{n-1}{2}.$$

Hence there is a Hamiltonian cycle in $\overline{G}'$ and there is a pairing of $[n/2]$ edges; that is, in G' there is a system π of $[n/2]$ pairs of vertices such that the vertices of any pair are not joined by an edge.

To a contraction of an edge (v_1, v_2) in $\overline{G}$ there corresponds in G an identification; that is, the replacement of the nonadjacent vertices v_1 and v_2 by a new vertex $\tilde{v}$ that is adjacent to just those vertices that are adjacent in G to both v_1 and v_2.

Suppose we are given an arbitrary system π of $[n/2]$ pairs of vertices of G, corresponding to a pairing in $\overline{G}'$. Identifying the vertices of each pair in G and rejecting the unpaired vertex if n is odd, we obtain a graph H_π. Among the H_π we choose the $H \doteq H_{\pi_0}$ with the fewest edges. Let $V(H) = \{v_i | i = 1, \dots, [n/2]\}$ and suppose that v_i is obtained by identifying v_{1i} and v_{2i} from $V(G)$. We observe that

$$(v_i, v_j) \in E(H) \Leftrightarrow \{(v_{1i}, v_{1j}), (v_{1i}, v_{2j}), (v_{2i}, v_{1j}), (v_{2i}, v_{2j})\} \subset E(G).$$

We define a *fork* on a pair of vertices x_1 and x_2 of $V(G)$ as a path (x_1, x_3, x_2) of length 2 joining x_1 and x_2 in G. We shall call (x_1, x_2) the *support* of this fork.

If (v_{1i}, v_{1j}), $(v_{2i}, v_{2j}) \notin E(G)$ (respectively (v_{1i}, v_{2j}), $(v_{2i}, v_{1j}) \notin E(G)$) for some $i \neq j$, i, $j \in \{1, 2, \ldots, [n/2]\}$, then the $(i, j, 1)$-transform (respectively, $(i, j, 2)$-transform) of H is the graph H^1_{ij} (respectively, H^2_{ij}) corresponding to the system π^1_{ij} (π^2_{ij}) that differs from π_0 only in the fact that the pairs $\{v_{1i}, v_{2i}\}$ and $\{v_{1j}, v_{2j}\}$ are replaced by the pairs $\{v_{1i}, v_{1j}\}$ and $\{v_{2i}, v_{2j}\}$ (respectively, by $\{v_{1i}, v_{2j}\}$ and $\{v_{2i}, v_{1j}\}$). Each of these transforms will be called a $(1, j)$-*transform.*

The edges of $E(G)$ hinder the existence of certain (i, j)-transforms. Since any edge can hinder no more than one transform, every vertex $v_i \in V(H)$ can occur in at least $n - 3 - 2(n(k+1)/3k - 1)$ transforms. Let $s_H(v_i) = r_i$ $(1 \leq i \leq [n/2])$. In the (i, j, l)-transform, $r_i + r_j$ edges of $E(H) \backslash E(H^l_{ij})$ are destroyed, but new edges may be formed. By virtue of the choice of H, for any $1 \leq i$, $j \leq [n/2]$ and $1 \leq l \leq 2$ we necessarily have $|E(H^l_{ij}) \backslash E(H)| \geq r_i + r_j$.

To each edge of $E(H^l_{ij}) \backslash E(H)$ there corresponds in G a pair of forks that has a common support from $\pi_0 \cap \pi^l_{ij}$. Any pair of forks can occur in only one transform in the creation of a "new" edge. Consequently, to each (i, j, l)-transform there correspond $x^l_{ij} \geq r_i + r_j$ pairs of forks, and any pair of such forks has a common support belonging to π_0. We associate any r_i of these pairs of forks with the vertex v_i, and any r_j of the remaining $x^l_{ij} - r_i$ pairs of forks with the vertex v_j. In all, to a vertex v_i $(1 \leq i \leq [n/2])$ there correspond no fewer than $r_i(n - (2n(k+1)/3k) - 1)$ pairs of forks. Moreover, to each edge $(v_i, v_j) \in E(H)$ there correspond the pairs of forks

$$\{(v_{1i}, v_{1j}, v_{2i}),\ (v_{1i}, v_{2j}, v_{2i})\}$$

and

$$\{(v_{1j}, v_{1i}, v_{2j}),\ (v_{1j}, v_{2i}, v_{2j})\},$$

which do not correspond to one vertex. Hence the total number of pairs of forks in G on pairs of vertices of π_0 is not less than

$$\sum_{i=1}^{[n/2]} r_i \left(\frac{n(k-2)}{3k} - 1 \right) + 2m(H) = \left(\frac{n(k-2)}{3k} - 1 \right) \cdot 2m(H) + 2m(H)$$
$$= 2m(H) \frac{n(k-2)}{3k}.$$

On the other hand, since $(v_{1j}, v_{2j}) \notin E(G')$ for any $1 \leq j \leq [n/2]$, there are no more than $3n/(k-1)(k-2)$ forks on $\{v_{1j}, v_{2j}\}$ and consequently no more than $\binom{3n/(k-1)(k-2)}{2}$ pairs of forks. Hence,

$$\frac{2m(H) \cdot n(k-2)}{3k} \leq \left[\frac{n}{2} \right] \cdot \frac{9n}{2(k-1)^2(k-2)^2} \cdot \left(n - \frac{(k-1)(k-2)}{3} \right),$$

and

$$2m(H) \leq \left[\frac{n}{2} \right] \cdot \frac{n-1}{2} \cdot \frac{27k}{(k-1)^2(k-2)^3}.$$

This proves the lemma.

LEMMA 8. *Let G be an n-vertex graph. Put $c = n^2/2m(G)$ and $z = \sigma(G)c/n$. Suppose that $c \geq 8.5z$ and $z \leq e^{\sqrt{\ln c}}/3e^{0.25}$. Then $\overline{G}$ can be contracted to an $[n/2]$-vertex graph $\overline{H}$ such that either $f(H;1) \geq f(G;1)$ or $\overline{H} = K_{[n/2]}$.*

PROOF. By Lemma 7, $\overline{G}$ can be contracted to an $[n/2]$-vertex graph $\overline{H}$ such that $E(H) = \varnothing$ or

$$c_1 \doteq \frac{n^2(H)}{2m(H)} \geq \left(\frac{c}{z}\right)^4 \cdot \frac{1}{27} \cdot \left(1 - \frac{z}{c}\right)^2 \cdot \left(1 - \frac{27}{c}\right)^3.$$

Suppose that $E(H) \neq \varnothing$. For $z/c \leq 1/8.5$ we have $(1 - z/c)^2(1 - 27/c)^3 > \frac{1}{3}$. Hence, $c_1 \geq (c/z)^4/81$ and

$$\begin{aligned} f(H;1) - f(G;1) &= \frac{[n/2](\sqrt{\ln c_1} - 1)}{2\sqrt{\ln[n/2]}} - \frac{n(\sqrt{\ln c} - 1)}{2\sqrt{\ln n}} \\ &\geq \frac{n}{2\sqrt{\ln n}}\left(\frac{1}{2}\sqrt{\ln c_1} - \sqrt{\ln c} + \frac{1}{2}\right) \geq \frac{n}{2\sqrt{\ln n}}\left(\sqrt{\ln(c/3z)} - \sqrt{\ln c} + \frac{1}{2}\right) \\ &= \frac{n}{2\sqrt{\ln n}} \cdot \frac{\ln c - \ln 3z - \ln c + \sqrt{\ln c} - 0.25}{\sqrt{\ln(c/3z)} + \sqrt{\ln c} - 1/2}. \end{aligned}$$

Thus, it is sufficient to prove the inequality $\ln 3z \leq \sqrt{\ln c} - 0.25$, which is equivalent to the condition $e^{\sqrt{\ln c}}/3e^{0.25} \geq z$ of our lemma.

LEMMA 9. *Let G be an n-vertex graph. Put $c = n^2/2m(G)$ and $z = \sigma(G)c/n$. Suppose that $c \geq 13$ and $z \leq e^{1.1\sqrt{\ln c}}/4$. Then $\overline{G}$ can be contracted to an $[n/2]$-vertex graph $\overline{H}$ such that either $f(H;1.1) \geq f(G;1.1)$ or $\overline{H} = K_{[n/2]}$.*

PROOF. By Lemma 7, $\overline{G}$ can be contracted to an $[n/2]$-vertex graph $\overline{H}$ such that either $E(H) = \varnothing$ or

$$c_1 \doteq \frac{n^2(H)}{2m(H)} \geq \left(\frac{c}{z}\right)^4 \cdot \frac{1}{27} \cdot \left(1 - \frac{z}{c}\right)^2 \cdot \left(1 - \frac{27}{c}\right).$$

Suppose that $E(H) \neq \varnothing$. For $c \geq 13$ we have $e^{1.1\sqrt{\ln c}}/4c \leq 1/8.9$, and for $c/z \geq 8.9$ we have

$$\left(1 - \frac{z}{c}\right)^2 \left(1 - \frac{27}{c}\right)^3 \geq \frac{27}{76}.$$

Consequently, $c_1 \geq (c/z)^4/76$ and

$$\begin{aligned} f(H;1.1) - f(G;1.1) &= \frac{[n/2](\sqrt{\ln c_1} - 1.1)}{2\sqrt{\ln[n/2]}} - \frac{n(\sqrt{\ln c} - 1.1)}{2\sqrt{\ln n}} \\ &\geq \frac{n}{2\sqrt{\ln n}}\left(\frac{1}{2}\sqrt{\ln c_1} - \sqrt{\ln c} + 0.55\right) \\ &= \frac{n}{2\sqrt{\ln n}} \cdot \frac{0.25\ln c_1 - (\sqrt{\ln c} - 0.55)^2}{0.5\sqrt{\ln c_1} + \sqrt{\ln c} - 0.55} \\ &\geq \frac{n}{2\sqrt{\ln n}} \cdot \frac{\ln c - \ln z - 0.25\ln 76 - \ln c + 1.1\sqrt{\ln c} - 0.3025}{0.5\sqrt{\ln c_1} + \sqrt{\ln c} - 0.55}. \end{aligned}$$

Now to prove the inequality $f(H;1.1) \geq f(G;1.1)$ it is sufficient to check that $\ln z \leq 1.1\sqrt{\ln c} - 0.25(1.21 + \ln 76)$. But since $\exp\left\{\frac{1}{4}(1.21 + \ln 76)\right\} < 4$, this inequality follows from the condition $z \leq e^{1.1\sqrt{\ln c}}/4$ of our lemma.

§4. Lower bounds for $\eta(\overline{G})$ expressed in terms of $f(G;\beta)$

LEMMA 10. *Let G be an n-vertex graph, and suppose that* $0 < 2m(G) \leq n^2/13$. *Then* $\eta(\overline{G}) \geq f(G;1.1)$.

PROOF. It is easy to verify the lemma for $n \leq 13$. Let G be a graph with the smallest number of vertices for which the lemma is false. Then $n \geq 14$. We consider five cases.

Case 1. Suppose that $0 < 2m(G) < n$. Then in G there is an isolated vertex v_1. Let $H = G\backslash\{v_1\}$. By the inductive hypothesis (IH), $\eta(\overline{H}) \geq f(H;1.1)$. But $\eta(\overline{G}) = \eta(\overline{H}) + 1 \geq f(H;1.1) + 1$, and by Lemma 2 we have $f(H;1.1) + 1 \geq f(G;1.1)$. For the rest of the lemma we assume that $c = n^2/2m(G)$ and $\sigma(G) = s_G(v_0) = zn/c$.

Case 2. Suppose that $2m(G) \geq n$ and $z \leq e^{1.1\sqrt{\ln c}}/4$. Since $c \leq n$ we have $f(G;1.1) < [n/2]$. Let H be the graph with $[n/2]$ vertices from the statement of Lemma 9. If $E(H) = \varnothing$, then $\eta(\overline{G}) \geq [n/2] \geq f(G;1.1)$. If $m(H) > 0$, then, by Lemma 7, $n^2(H)/2m(H) > 13$, and by Lemma 9, $f(H;1.1) \geq f(G;1.1)$. Consequently, by the IH, $\eta(\overline{H}) \geq f(H;1.1) \geq f(G;1.1)$.

Case 3. Suppose that $2m(G) \geq n$ and

$$s_G(v_0) + \sigma(G\backslash N(v_0)) \geq n \cdot (1 + \ln c - 1.1\sqrt{\ln c})/c.$$

Suppose also that $v_1 \in V(G)\backslash N(v_0)$ and $s_{G\backslash N(v_0)}(v_1) = \sigma(G\backslash N(v_0))$. Then identifying v_0 and v_1 we obtain a graph G' with $n(G') = n - 1$ and

$$m(G') = m(G) - s_G(v_0) - s_{G\backslash N(v_0)}(v_1) \leq m(G) - 2m(G)/n(1 + \ln c - 1.1\sqrt{\ln c}).$$

It is easy to check that for $c \geq 13$ we have $2m(G') \leq (n-1)^2/13$, and, by the IH and Lemma 3,

$$\eta(\overline{G}) \geq \eta(\overline{G}') \geq f(G';1.1) \geq f(G;1.1).$$

Case 4. The conditions of Cases 1–3 are not satisfied and $c \geq e^{2.89}$. Consider $G' \doteq G\backslash N(v)$. Then $n(G') = ((c - z)/c) \cdot n$ and

$$\sigma(G') < (n/c)(1 + \ln c - 1.1\sqrt{\ln c} - e^{1.1\sqrt{\ln c}}/4).$$

Let us put $k = n(G')/\sigma(G')$. Putting $x = \sqrt{\ln c}$ in Lemma 6, we have $k \geq c$. Let us apply Lemma 7 to G'. If $n(G')$ is odd, we take v_0 for a vertex that does not occur in one of the identified pairs. Taking account of the fact that v_0 is an isolated vertex in G', we find that $\overline{G}$ can be contracted to an $[(n(G') + 1)/2]$-vertex graph $\overline{H}$ such that

$$\begin{aligned} 2m(H) &\leq \frac{n^2(H) \cdot 27k}{(k-1)^2(k-2)^3} \leq \frac{n^2(H) \cdot 27c}{(c-1)^2(c-2)^3} \\ &\leq \frac{n^2(H) \cdot 27}{c^4}\left(1 - \frac{1}{17.99}\right)^{-2}\left(1 - \frac{2}{17.99}\right)^{-3} \leq \frac{e^4}{c^4} \cdot n^2(H). \end{aligned}$$

If $m(H) = 0$, then

$$\eta(\overline{G}) \geq \eta(\overline{G}') \geq n(H) \geq \tfrac{1}{2}n(1 - z/c) \geq \tfrac{1}{2}n(1 - \ln c/c).$$

Since $e^{2.89} \leq c \leq n$, we have

$$\frac{n}{2}\left(1 - \frac{\ln c}{c}\right) \geq \frac{n}{2}\left(1 - \frac{1}{\sqrt{\ln c}}\right) \geq \frac{n(\sqrt{\ln c} - 1)}{2\sqrt{\ln n}} \geq f(G; 1.1).$$

Suppose that $m(H) > 0$. We show that $f(H; 1.1) \geq f(G; 1.1)$. In fact,

$$\begin{aligned}
f(H;1.1) - f(G;1.1) &\geq \frac{n(1 - z/c)(\sqrt{\ln(c^4/e^4)} - 1.1)}{2 \cdot 2\sqrt{\ln(n(1 - z/c)/2)}} - \frac{n(\sqrt{\ln c} - 1.1)}{2\sqrt{\ln n}} \\
&\geq \frac{n}{2c\sqrt{\ln n}}((c - z)(\sqrt{\ln c - 1} - 0.55) - c(\sqrt{\ln c} - 1.1)) \\
&\geq \frac{n}{2c\sqrt{\ln n}}((c - \ln c)\sqrt{\ln c - 1} + 0.55c + 0.55 \ln c - c\sqrt{\ln c}) \\
&\geq \frac{n}{2c\sqrt{\ln n}}\left(c \cdot \frac{-1}{\sqrt{\ln c - 1} + \sqrt{\ln c}} + 0.55c - \ln c(\sqrt{\ln c - 1} - 0.55)\right).
\end{aligned}$$

Since $c \geq e^{2.89}$, we have

$$\begin{aligned}
f(H;1.1) - f(G;1.1) &\geq \frac{n}{2c\sqrt{\ln n}}\left(\frac{-c}{1.37 + 1.7} + 0.55c - \ln^{3/2} c + 0.55 \cdot 2.89\right) \\
&\geq \frac{n}{2c\sqrt{\ln n}}(0.224c + 1.58 - \ln^{3/2} c) > 0.
\end{aligned}$$

From the form of the function f and the fact that $f(H; 1.1) \geq f(G; 1.1)$ it follows that $c_1 \doteq n^2(H)/2m(H) > c$. Then, by the IH, $\eta(\overline{H}) \geq f(H; 1.1) \geq f(G; 1.1)$.

Case 5. The conditions of Cases 1–3 are not satisfied and $c \in [13, e^{2.89}]$. Let us consider $G' \doteq G \backslash N(v_0)$. As in Case 4, $n(G') = ((c - z)/c)n$, and

$$m(G') \leq \frac{n(G') \cdot \sigma(G')}{2} \leq \frac{(c - z) \cdot n^2}{2c^2}(1 + \ln c - 1.1\sqrt{\ln c} - z).$$

If $m(G') = 0$, then $\eta(\overline{G}) \geq n(c - z)/c \geq f(G; 1.1)$. Suppose that $m(G') > 0$. Then

$$c_1 = n^2(G')/2m(G) \geq (c - z)(1 + \ln c - 1.1\sqrt{\ln c} - z)$$

We show that $f(G'; 1.1) \geq f(G; 1.1)$. In fact,

$$\begin{aligned}
f(G';1.1) - f(G;1.1) &\geq \frac{c - z}{c} \cdot \frac{n\left(\sqrt{\ln \dfrac{c - z}{1 + \ln c - 1.1\sqrt{\ln c} - z}} - 1.1\right)}{2\sqrt{\ln n}} \\
&\quad - \frac{n(\sqrt{\ln c} - 1.1)}{2\sqrt{\ln n}} \\
&= \frac{n}{2c\sqrt{\ln n}} \cdot \varphi(c; z; \beta),
\end{aligned}$$

where $\varphi(c; z; \beta)$ was defined at the beginning of §2.

By Lemmas 4 and 5, to check that $\varphi(c;z;\beta) \geq 0$ for $c \in [13, e^{2.89}]$ and $z \geq e^{1.1\sqrt{\ln c}}/4$ it is sufficient to verify it for $c = e^{2.89}$ and any $z \leq e^{1.1\sqrt{\ln 13}}/4$. In the table at the end of the paper it is verified for $z = 1.45$.

Thus, $f(G';1.1) \geq f(G;1.1)$ and $n(G') < n$. From the definition of f we have $n^2(G')/2m(G') > c \geq 13$, and by the IH we have $\eta(\overline{G}') \geq f(G';1.1) \geq f(G;1.1)$. This proves the lemma.

LEMMA 11. *Let G be an n-vertex graph and suppose that $0 < 2m(G) \leq n^2/6.2$. Then $\eta(\overline{G}) \geq 5f(G;1)/6$.*

PROOF. For $n \leq 7$ the lemma is obvious. Let G be the smallest counterexample. If $0 < 2m(G) < n$, then there is an isolated vertex v_1 in G. Since $2m(G) \leq n-1$, and $n \geq 8$, by the IH we have

$$\eta(\overline{G}\backslash\{v_1\}) \geq 5f(G\backslash\{v_1\};1)/6.$$

But $\eta(\overline{G}) = \eta(\overline{G}\backslash\{v_1\}) + 1$, and $5f(G\backslash\{v_1\};1)/6 + 1 \geq 5f(G)/6$ by Lemma 2. Thus, $2m(G) \geq n$.

If $2m(G) \leq n^2/13$, then $5f(G;1)/6 \leq f(G;1.1)$, and taking account of Lemma 10 we have $\eta(\overline{G}) \geq 5f(G;1)/6$.

We may thus assume that $c \in [6.2, 13]$, where $c = n^2/2m(G)$ and $\sigma(G) = s_G(v_0) = zn/c$. By Lemma 8, for $c \geq 9.4$ we necessarily have $z \geq e^{\sqrt{\ln c}}/3e^{0.25}$. Otherwise $\overline{G}$ could be contracted to a graph H with $f(H;1) \geq f(G;1)$. Let us put $G' = G\backslash N(v_0)$. Then $n(G') = (c-z)n/c$. As in Case 3 of Lemma 10, it is easy to check that it suffices to consider the case

$$zn/c + \sigma(G') < (1 + \ln c - \sqrt{\ln c})n/c.$$

Then

$$m(G') \leq \frac{n(G')\cdot\sigma(G')}{2} \leq \frac{n^2(c-z)(1+\ln c - \sqrt{\ln c} - z)}{2c^2}.$$

If $m(G') = 0$, then $\eta(\overline{G}) \geq (c-z)n/c > 5f(G;1)/6$, since $z < \ln c$. Suppose that $m(G') > 0$. Then

$$\begin{aligned}
&\frac{5}{6}f(G';1) - \frac{5}{6}f(G;1)\\
&\geq \frac{5}{6}\left(\frac{n(c-z)\left(\sqrt{\ln\dfrac{c-z}{1+\ln c\sqrt{\ln c}-z}}-1\right)}{2c\sqrt{\ln\dfrac{n(c-z)}{c}}} - \frac{n(\sqrt{\ln c}-1)}{2\sqrt{\ln n}}\right)\\
&\geq \frac{5n}{12c\sqrt{\ln n}}\cdot\varphi(c;z;1).
\end{aligned}$$

By Lemmas 4 and 5, to prove that $\varphi(c;z;1) \geq 0$ for $c \in [a,b]$ and for $e^{\sqrt{\ln c}}/3e^{0.25} \leq z < 1 + \ln c - \sqrt{\ln c}$ it is sufficient to verify it for $c = b$ and any $z \leq \exp\{\sqrt{\ln a} - 0.25\}/3$. Hence, the corresponding calculations in the table confirm the truth of this inequality on the intervals [9.4; 10.1], [10.1; 10.67], [10.67; 11.2], and [12; 13]. Similarly we consider $c \in [6.2; 9.4]$. We merely need to take account of the fact that, by definition, $z \geq 1$. Therefore, $f(G'; 1) \geq f(G; 1)$ and by the IH we have $\eta(\overline{G}') \geq 5f(G'; 1)/6$. Hence, $\eta(\overline{G}) \geq \eta(\overline{G}') \geq 5f(G; 1)/6$.

As in Lemma 11, by using the table we can prove the following facts.

LEMMA 12. *Let G be an n-vertex graph, and suppose that $0 < 2m(G) \leq n^2/4.5$. Then $\eta(\overline{G}) \geq (35/57)f(G; 7/8)$.*

LEMMA 13. *Let G be an n-vertex graph, and suppose that $0 < 2m(G) \leq n^2/3.15$. Then $\eta(\overline{G}) \geq 0.4616 f(G; 0.75)$.*

LEMMA 14. *Let G be an n-vertex graph, and suppose that $0 < 2m(G) \leq n^2/2.55$. Then $\eta(\overline{G}) \geq 0.3663 f(G; 2/3)$.*

LEMMA 15. *Let G be an n-vertex graph, and suppose that $0 < 2m(G) \leq n^2/2$. Then $\eta(\overline{G}) \geq 0.299 f(G; 0.6)$.*

§5. Proof of the theorem

From the results of Mader [8] and Tashkinov (course paper, Novosibirsk State University, 1976) for $k \leq 5$, it follows that the theorem is true for $k \leq 100$. Thus, suppose that $k > 100$ and that $G \in \mathcal{E}_k$ is a counterexample to the theorem with the smallest number of edges for the given k. If $n(G) = k$, then $m(G) > k^2 - \binom{k+1}{2} = \binom{k}{2}$, which is impossible. Hence, $n(G) > k$. By the choice of G,

$$m(G) = k \cdot n(G) - \binom{k+1}{2} + 1.$$

Under a contraction of the edge $(u, v) \in E(G)$ the number of edges in the resulting graph G' is less than $|E(G) \backslash \{(u, v)\}|$ by the number of vertices in G adjacent to both u and v, that is, by the number of triangles containing the edge (u, v). Hence, if (u, v) occurs in fewer than k triangles, we have

$$m(G') \geq m(G) - (k-1) - 1 = k \cdot n(G) - \binom{k+1}{2} + 1 - k = k \cdot n(G') - \binom{k+1}{2} + 1.$$

By the IH, we have $\eta(G') \geq 0.064k/\sqrt{\ln k} + 1$, and $\eta(G) \geq \eta(G')$. Consequently, every edge of $E(G)$ occurs in no fewer than k triangles. Thus, for any vertex v the degree of each vertex $w \in N_G(v)$ in the graph $G(N(v))$ is not less than k. Suppose that $\min_{v \in V(G)} s_G(v) = s_G(v_0) = \alpha k$. By what we stated above, $\alpha > 1$. Since $m(G)/n(G) < k$, we have $\alpha < 2$. Let us put $G_0 = G(N(v_0))$. Then $\sigma(\overline{G}_0) \leq \alpha k - 1 - k < \alpha k/2$. Consequently, by Lemma 15,

$$\eta(G_0) = 0.299 \frac{\sqrt{\ln(\alpha/(\alpha-1))} - 0.6}{2\sqrt{\ln \alpha \cdot k}} \cdot \alpha \cdot k.$$

Let $\Psi(\alpha) \doteq \alpha(\sqrt{\ln(\alpha/(\alpha-1))} - 0.6)$; then

$$\Psi'(\alpha) = \frac{2\ln(\alpha/(\alpha-1)) - 1/(\alpha-1) - 1.2\sqrt{\ln(\alpha/(\alpha-1))}}{2\sqrt{\ln(\alpha/(\alpha-1))}}.$$

Putting $\gamma \doteq \alpha/(\alpha-1)$, we have $\gamma > 2$ and

$$\Psi'(\alpha) = (2\ln\gamma - \gamma + 1 - 1.2\sqrt{\ln\gamma})/2\sqrt{\ln\gamma}.$$

By Lemma 1, $\Psi'(\alpha) \leq 0$. Hence, taking account of the fact that $1/\sqrt{k}\ln\alpha$ decreases as α increases, we have

$$\begin{aligned}\eta(g_0) &\geq 0.299 \cdot \frac{\sqrt{\ln(2/(2-1))} - 0.6}{2\sqrt{\ln 2k}} \cdot 2k \\ &\geq 0.299 \cdot \frac{0.832 - 0.6}{\sqrt{\ln 2k}} \cdot k \geq 0.0692 \frac{k}{\sqrt{\ln 2k}} \\ &\geq \frac{k}{\sqrt{\ln k}} \cdot \frac{0.0692}{\sqrt{1 + 1/\log_2 100}} \geq 0.064 \cdot \frac{k}{\sqrt{\ln k}}.\end{aligned}$$

but since v_0 is adjacent to all the vertices of $V(G_0)$, we have $\eta(G) \geq \eta(G_0) + 1$. This proves the theorem.

BIBLIOGRAPHY

1. K. Wagner, *Über eine Eigenschaft der ebenen Komplexe*, Math. Ann. **114** (1937), 570–590.

2. H. Hadwiger, *Über eine Klassifikation der Streckenkomplexe*, Vierteljahrsschrift Naturforsch. Ges. Zürich **88** (1943), 133–142.

3. K. Wagner, *Beweis einer Abschwächung der Hadwiger-Vermutung*, Math. Ann. **153** (1964), 139–141.

4. Paul Erdős and Joel Spencer, *Probabilistic methods in combinatorics*, Academic Press, 1974.

5. A. A. Zykov, *On the number of edges of a graph with Hadwiger number not greater than* 4, Prikl. Mat. i Programmirovanie Vyp. 7 (1972), 52–55. (Russian)

6. Bohdan Zelinka, *On some graph-theoretical problems of V. G. Vising*, Časopio Pěst. Mat. **98** (1973), 56–66.

7. Zevi Miller, *Contractions of graphs: a theorem of Ore and an extremal problem*, Discrete Math. **21** (1978), 261–272.

8. W. Mader, *Homomorphiesätze für Graphen*, Math. Ann. **178** (1968), 154–168.

9. A. D. Korshunov, *On the chromatic number of n-vertex graphs*, Metody Diskret. Anal. Vyp. **35** (1980), 15–44. (Russian)

Appendix

PROOF OF LEMMA 1. The assertion of the lemma is true for $x = 2$, and for $x \geq 2$ we have

$$(x - 1 - 2\ln x + 2\beta\sqrt{\ln x})'_x = 1 - \frac{2}{x} + \frac{\beta}{x \cdot \sqrt{\ln x}} > 0.$$

PROOF OF LEMMA 2. Let us put $c = n^2/2m$. We note that, by hypothesis, $c \geq n$. Then

$$\begin{aligned}
&f(n-1;m;\beta) - f(n;m;\beta) + 1 \\
&\quad = \frac{n-1}{2\sqrt{\ln(n-1)}}\left(\sqrt{\ln\frac{(n-1)^2}{2m}} - \beta\right) - \frac{n(\sqrt{\ln c} - \beta)}{2\sqrt{\ln n}} + 1 \\
&\quad \geq \frac{n}{2\sqrt{\ln n}}\left[\left(1-\frac{1}{n}\right)(\sqrt{\ln c - 2/n} - \beta) - \sqrt{\ln c} + \beta + \frac{2}{n}\sqrt{\ln n}\right] \\
&\quad \geq \frac{n}{2\sqrt{\ln n}}\left[\left(1-\frac{1}{n}\right)\left(\sqrt{\ln c} - \frac{1}{n} - \beta\right) - \sqrt{\ln c} + \beta + \frac{2}{n}\sqrt{\ln n}\right] \\
&\quad \geq \frac{1}{2\sqrt{\ln n}} \cdot \left(\sqrt{\ln c} - \frac{\ln c}{n} - \frac{1}{n} + \frac{1}{n^2} - \beta + \frac{\beta}{n} - \sqrt{\ln c} + \beta + \frac{2}{n}\sqrt{\ln n}\right) \\
&\quad \geq \frac{1}{2\sqrt{\ln n}}(-\sqrt{\ln c} - 1 + \beta + 2\sqrt{\ln n}).
\end{aligned}$$

Since $\ln c < 2\ln n$, $n \geq 4$, and $\beta \geq 0.6$, we have

$$-\sqrt{\ln c} - 1 + \beta + 2\sqrt{\ln n} \geq (2 - \sqrt{2})\sqrt{\ln n} - 0.4 > 0.$$

PROOF OF LEMMA 3. Let us put $c = n^2/2m$, $c_1 = (n-1)^2/2(m - 2m\alpha/n)$. Then

$$\begin{aligned}
&f(n-1; m - 2m\alpha/n; \beta) - f(n;m;\beta) \\
&\quad \geq \frac{1}{2\sqrt{\ln n}}[(n-1)(\sqrt{\ln c_1} - \beta) - n(\sqrt{\ln c} - \beta)].
\end{aligned}$$

Hence it is sufficient to show that $(n-1)\sqrt{\ln c_1} \geq n\sqrt{\ln c} - \beta$. From the inequality for α in the condition of the lemma it follows that

$$2\ln c - 2\beta\sqrt{\ln c} \leq 2(\alpha - 1) < 2(\alpha-1)\cdot(n-1)^2/(n(n-2)).$$

Since $x < \ln(1-x)^{-1}$ for any $x \in (0,1)$, we have

$$\begin{aligned}
2(\alpha-1)\cdot\frac{(n-1)^2}{n(n-2)} &\leq \frac{(n-1)^2}{n}\cdot\ln\left(1 - \frac{2(\alpha-1)}{n-2}\right)^{-1} = \frac{(n-1)^2}{n}\cdot\ln\frac{n-2}{n-2\alpha} \\
&\leq \frac{(n-1)^2}{n}\cdot\ln\frac{(1-1/n)^2}{1-2\alpha/n} = \frac{(n-1)^2}{n}\left(\ln\frac{n^2(1-1/n)^2}{2m(1-2\alpha/n)} - \ln\frac{n^2}{2m}\right) \\
&= \frac{(n-1)^2}{n}(\ln c_1 - \ln c),
\end{aligned}$$

that is,

$$(n-1)^2(\ln c_1 - \ln c) \geq 2n\ln c - 2\beta n\sqrt{\ln c}.$$

Consequently,

$$\begin{aligned}
(n-1)^2\ln c_1 &\geq n^2\ln c + \ln c - 2\beta n\sqrt{\ln c} \\
&= (n\sqrt{\ln c} - \beta)^2 + (\ln c - \beta^2) > (n\sqrt{\ln c} - \beta)^2,
\end{aligned}$$

as required.

PROOF OF LEMMA 4. Putting $\alpha \doteq 1+\ln c-\beta\sqrt{\ln c}$ and $y \doteq (c-z)/(\alpha-z)$, we have

$$\varphi'(c;z;\beta) = -\sqrt{\ln\frac{c-z}{\alpha-z}} + \beta - (c-z)\frac{\dfrac{1}{c-z}-\dfrac{1}{\alpha-z}}{2\sqrt{\ln\dfrac{c-z}{\alpha-z}}}$$

$$= \frac{y-1-2\ln y+2\beta\sqrt{\ln y}}{2\sqrt{\ln y}}.$$

Since

$$y = \frac{c-z}{1+\ln c-\beta\sqrt{\ln c}-z} \geq \frac{c-1}{\ln c-\beta\sqrt{\ln c}},$$

by Lemma 1 for c, z and β that satisfy the condition of our lemma we have $y(c;z;\beta) \geq 2$. Hence, by Lemma 1 we have $\varphi'_z(c;z;\beta) > 0$.

PROOF OF LEMMA 5. Putting

$$u \doteq \frac{c-z}{c(1+\ln c-\beta\sqrt{\ln c}-z)},$$

we have

$$\varphi'_c(c;z;\beta) = \sqrt{\ln\frac{c-z}{1+\ln c-\beta\sqrt{\ln c}-z}}$$

$$+\left(1-\frac{(c-z)(1-\beta/2\sqrt{\ln c})}{c(1+\ln c-\beta\sqrt{\ln c}-z)}\right)\Big/2\sqrt{\ln\frac{c-z}{1+\ln c-\beta\sqrt{\ln c}-z}}$$

$$-\sqrt{\ln c}-1/2\sqrt{\ln c}$$

$$= \sqrt{\ln cu}+\left(1-u\left(1-\frac{\beta}{2\sqrt{\ln c}}\right)\right)\Big/2\sqrt{\ln cu}-\sqrt{\ln c}-\frac{1}{2\sqrt{\ln c}}$$

$$= (\sqrt{\ln cu}-\sqrt{\ln c})\left(1-\frac{1}{2\sqrt{\ln c}\sqrt{\ln cu}}\right)-u\left(1-\frac{\beta}{2\sqrt{\ln c}}\right)\Big/2\sqrt{\ln cu}$$

$$= \frac{\ln u}{\sqrt{\ln cu}+\sqrt{\ln c}}\left(1-\frac{1}{2\sqrt{\ln c}\cdot\sqrt{\ln cu}}\right)-u\left(1-\frac{\beta}{2\sqrt{\ln c}}\right)\Big/2\sqrt{\ln cu}.$$

By Lemma 1, $cu > 2$. Hence, $1-1/2\sqrt{\ln c}\sqrt{lncu} > 0$. If $\ln u \leq 0$, then $\varphi'_c(c;z;\beta) < 0$. Suppose that $u > 1$. Then $\varphi'_c(c;z;\beta) < 0$ is equivalent to

$$\frac{u}{\ln u}\left(1-\frac{\beta}{2\sqrt{\ln c}}\right) > \frac{2\sqrt{\ln cu}-1/\sqrt{\ln c}}{\sqrt{\ln cu}+\sqrt{\ln c}}. \qquad (**)$$

For any $u > 1$ we have $u/\ln u \geq e$. By the condition of the lemma, $1-\beta/2\sqrt{\ln c} > 1/2$. Hence, if the right side of $(**)$ does not exceed $4/3 < e/2$, then $(**)$ is true. Suppose that

$$2\sqrt{\ln cu}-1/\sqrt{\ln c} > \frac{4}{3}(\sqrt{\ln c}+\sqrt{\ln cu}).$$

Then $\sqrt{\ln cu} > 2\sqrt{\ln c}/3+3/2\sqrt{\ln c}$, $\ln cu > 4\ln c/9+6+9/4\ln c$, and $\ln u > 6$. But $u/\ln u > 4$ for $u > e^6$, and since the right-hand side of $(**)$ does not exceed 2, $(**)$ is true.

PROOF OF LEMMA 6. For $x \geq 1.7$ we have that $g(x) \geq h(x) \doteq e^{1.1x}/4 + 1.1x - x^2 - 0.1$, and $h'(x) = 1.1e^{1.1x}/4 + 1.1 - 2x$ is a convex downwards function. Hence, h has no more than one minimum. We observe that

$$h'(2.3) = 1.1e^{2.53}/4 + 1.1 - 4.6 \leq 1.1 \cdot 12.56/4 - 3.5 < 3.46 - 3.5 < 0;$$
$$h'(2.34) = 1.1e^{2.574}/4 + 1.1 - 4.68 \geq 1.1 \cdot 13.118/4 - 3.58 > 3.6 - 3.58 > 0.$$

Consequently, it is sufficient to verify that $h(x) > 0$ for $x = 1.7$ and $x \in [2.3, 2.34]$. But $h(1.7) = e^{1.87}/4 + 1.87 - 2.89 - 0.1 \geq 6.488/4 - 1.12 > 0$. If $x \in [2.3, 2.34]$, then

$$\begin{aligned} h(x) &\geq e^{1.1 \cdot 2.3}/4 - (2.34^2 - 1.1 \cdot 2.34) - 0.1 \geq 12.552/4 - 2.34 \cdot 1.24 - 0.1 \\ &= 3.138 - 2.9016 - 0.1 > 0. \end{aligned}$$

Translated by E. J. F. PRIMROSE

Table

β	z	c	$\sqrt{\ln c} \le$	$1+\ln c - \beta\sqrt{\ln c} - z \le$	$\frac{c-z}{1+\ln c - \beta\sqrt{\ln c} - z} \ge$	$\sqrt{\ln \frac{c-z}{1+\ln c - \beta\sqrt{\ln c} - z}} \ge$	$\varphi(c;z;\beta) \ge$
2/3	1	3.15	1.072	0.4346	4.947	1.264	0.007
3/4	1	4.15	1.193	0.5285	5.96	1.336	0.0074
7/8	1	6.2	1.351	0.6431	8.085	1.445	0.012
1	1	9.4	1.497	0.7441	11.288	1.5567	0.004
1	1.1	10.1	1.521	0.69246	12.99	1.601	0.14
1	1.17	10.67	1.539	0.66	14.3939	1.633	0.26
1	1.2	11.2	1.555	0.6631	15.08	1.647	0.2
1	1.227	12	1.577	0.683	15.77	1.66	0.28
1	1.2545	13	1.602	0.71	16.54	1.674	0.09
1.1	1.45	17.99	1.7	0.57	29.017	1.835	1.3
0.6	1	2.55	0.968	0.3561	4.35	1.212	0.01

Amer. Math. Soc. Transl.
(2) Vol. 132, 1986

An Upper Bound on the Number of Edges of a Hypergraph with a Given Diameter

O. G. RUDENSKAYA

An upper bound on the number of edges of a connected simple hypergraph with a given diameter is established, and the class of extremal hypergraphs is described; an analogous description is obtained for the class of uniform connected simple hypergraphs with a given diameter and the maximum number of edges.

We use the terminology from [1]–[4].

Let H be an n-vertex connected hypergraph with diameter d, x the "initial" vertex of some chain of length d in H connecting two vertices which are distance d apart, and for $i = 0, \ldots, d$ let $\langle a_i \rangle$ be the set of vertices in H which are at a distance i from x and $a_i = |\langle a_i \rangle|$ (we will consider the distance from a vertex to itself to be 0).

We write H schematically as

$$\langle a_0 \rangle - \langle a_1 \rangle - \cdots - \langle a_{d-1} \rangle - \langle a_d \rangle, \tag{1}$$

where

$$a_0 = 1, \qquad a_i \geq 1 \quad \text{for } i = 1, \ldots, d, \qquad a_0 + a_1 + \cdots + a_d = n. \tag{2}$$

Clearly an edge in H is incident with only the vertices from $\langle a_{i-1} \rangle \cup \langle a_i \rangle$ for some $i = 1, \ldots, d$.

§**I.** We denote an n-vertex simple hypergraph with diameter d by $H(n, d)$.

By analogy with Theorem 2 of [3] it is easy to prove that $H(n, d)$ of the form (1) for $d \geq 2$ is diameter-critical (d.c.) if and only if its edges are all the possible combinations of $a_{i-1} + a_i$ vertices for j, $i = 1, \ldots, d$, $j = 1, \ldots, a_{a-1} + a_i$ and $a_0 = a_d = 1$.

1980 *Mathematics Subject Classification* (1985 *Revision*). Primary 05C65; Secondary 05C35.

Translation of Mat. Issled. Vyp. 66 (1982), 141–147; MR **83g**:05060.

Clearly the number of edges such a d.c. hypergraph is equal to

$$1+\sum_{i=1}^{d}\sum_{j=1}^{a_{i-1}+a_i}\binom{a_{i-1}+a_i}{j}-\binom{a_{i-1}}{j}. \tag{3}$$

We call a hypergraph of the form (1) which differs from a d.c. hypergraph only in that a_0 and a_d may be greater than 1 *almost diametrically critical* (a.d.c.).

With the help of the formula

$$\sum_{j=0}^{k}\binom{k}{j}=2^k, \tag{4}$$

where k is a natural number, and (3) it is easy to prove

LEMMA 1. *The number of edges in an a.d.c. hypergraph* $\langle 1\rangle$–$\langle a_1\rangle$–$\langle a_2\rangle$–$\langle a_3\rangle$ *is not greater than (less than if* $a_1, a_2 > 1$*) the number of edges in an a.d.c. hypergraph* $\langle 1\rangle$–$\langle 1\rangle$–$\langle a_2+a_1-1\rangle$–$\langle a_3\rangle$.

THEOREM 1. *If m is the number of edges in $H(n,d)$, then*

$$m\le\begin{cases}2^n-1 & \text{for } d=1,\\ 3\cdot 2^{n-d}+2d-5 & \text{for } d\ge 2.\end{cases} \tag{5}$$

PROOF. For $d=1$, clearly $m\le\sum_{i=1}^{n}\binom{n}{i}=2^n-1$.

For $d\ge 2$ we prove the theorem by induction.

Suppose that $d=2$; the bound (5) is true, since the number of edges in $\langle 1\rangle$–$\langle a_1\rangle$–$\langle a_2\rangle$ is not greater than the number of edges in a d.c. $\langle 1\rangle$–$\langle n-2\rangle$–$\langle 1\rangle$, which is equal to

$$2\sum_{i=2}^{n-1}\binom{n-1}{i}-\sum_{i=2}^{n-2}\binom{n-2}{i}=3\cdot 2^{n-2}-1=3\cdot 2^{n-d}+2d-5.$$

We suppose that the bound holds for $d=k-1$ and for any $n>k-1$, and prove it for $d=k$ and any $n>k$.

By Lemma 1, for $H(n,k)$ of the form $\langle 1\rangle$–$\langle a_1\rangle$–$\langle a_2\rangle$–$\langle a_3\rangle$–$\cdots$–$\langle a_{k-1}\rangle$–$\langle a_k\rangle$ with m edges there is a d.c. $\langle 1\rangle$–$\langle 1\rangle$–$\langle a_2+a_1-1\rangle$ – $\langle a_3\rangle$–$\cdots$–$\langle a_{k-1}\rangle$–$\langle a_k\rangle$ with $m_1\ge m$ edges.

By the induction hypothesis the number of edges in $H(n-1,k-1)$ of the form $\langle 1\rangle$–$\langle a_2+a_1-1\rangle$–$\langle a_3\rangle$–$\cdots$–$\langle a_{k-1}\rangle$–$\langle a_k\rangle$ is not greater than

$$3\cdot 2^{n-1-(k-1)}+2(k-1)-5=3\cdot 2^{n-k}+2k-7,$$

whence $m_1\le 3\cdot 2^{n-k}+2k-5$. The theorem is proved.

With the help of (4) it is easy to obtain

LEMMA 2.

$$\alpha=\sum_{i=2}^{a_1+1}\binom{a_1+1}{i}+\sum_{i=2}^{1+a_3}\binom{1+a_3}{i}-1-\sum_{i=2}^{a_1+a_3}\binom{a_1+a_3}{i}\le 2,$$

where $\alpha=2$ *if and only if* $a_1=1$ *or* $a_3=1$.

THEOREM 2. *A d.c. $H(n,d)$ of the form* (1), *where $d \geq 2$, is maximal* (*max. d.c.*) *if and only if there exists t, $1 \leq t \leq d-1$, such that*

$$a_i = \begin{cases} n-d & \text{for } i = t, \\ 1 & \text{for } i \neq t, \end{cases} \tag{6}$$

where $i = 0, \ldots, d$.

PROOF. The sufficiency is established by using formula (3) to count the number of edges in $H(n,d)$ of the form (1), with the a_i as given in (6).

The necessity is clearly satisfied for $d = 2$. We suppose it to be true for $d = k-1$ and prove it for $d = k$.

If $H' = H(n,k)$ is a max.d.c. hypergraph of the form of (1), then either $a_1 = 1$ or $a_2 = 1$: otherwise by Lemma 1 there would be an $H(n,k)$ with a greater number of edges.

Suppose that $a_1 = 1$. Clearly H' is max.d.c. if and only if $H(n-1, k-1)$ of the form $\langle 1\rangle$–$\langle a_2\rangle$–$\cdots$–$\langle a_k\rangle$ is max.d.c; that is, if and only if there exists t, $2 \leq t \leq k-1$, such that

$$a_i = \begin{cases} n-1-(k-1) = n-k & \text{for } i = t, \\ 1 & \text{for } i \neq t, \end{cases} \tag{7}$$

where $i = 2, \ldots, k$.

Suppose that $a_1 > 1$ and $a_2 = 1$. Consider a d.c. $H(n-1, k-1)$ of the form

$$\langle 1\rangle\text{–}\langle a_1\rangle\text{–}\langle a_3\rangle\text{–}\langle a_4\rangle\text{–}\cdots\text{–}\langle a_k\rangle, \tag{8}$$

obtained from H' by deleting $\langle a_2\rangle$. Clearly the number m of edges of the hypergraph H' is related to the number m_1 of edges of the hypergraph (8) by the relation $m = m_1 + \alpha$, where α is the same one as in Lemma 2.

Clearly m attains its maximum value if and only if 1) the hypergraph (8) is max.d.c., and 2) the number $\alpha = 2$ (if it is possible to satisfy conditions 1) and 2) simultaneously).

By the induction hypothesis, since the hypergraph (8) is max.d.c., there exists t, $1 \leq t \leq k-1$, $t \neq 2$, such that the a_i satisfy (7), where $i = 1, 3, \ldots, k$; thus, since $a_1 > 1$, we have $t = 1$. Since $\alpha = 2$, $a_3 = 1$ by Lemma 2. The theorem is proved.

§**II.** We now consider n-vertex uniform (of rank s) connected simple hypergraphs with diameter d; we denote any such hypergraph by $H(n,s,d)$.

Clearly, for $H(n,s,d)$ of the form (1) we have

$$a_{i-1} + a_i \geq s, \qquad i = 1, \ldots, d, \tag{9}$$

as well as (2).

It is known [3] that $H(n,s,d)$ of the form of (1) is d.c. if and only if its edges are all the possible combinations of $a_{i+1} + a_i$ vertices for $s, i = 1, \ldots, d$.

It follows that the number of edges of such a d.c. hypergraph is equal to

$$\binom{a_0}{s} + \sum_{i=1}^{d} \binom{a_{i-1}+a_i}{s} - \binom{a_{i-1}}{s}. \tag{10}$$

LEMMA 3. *If $a_0 \geq s-1$ and $a_3 \leq s-1$, then the number m_1 of edges in an a.d.c. $\langle a_0\rangle$–$\langle a_1\rangle$–$\langle a_2\rangle$–$\langle a_3\rangle$ is not greater than (less than, if $s \geq 3$, $a_0 + a_1 > s$, and $a_2 + a_3 > s$) the number m_2 of edges in an a.d.c. $\langle a_0\rangle$–$\langle a_1 + a_2 + a_3 - s\rangle$ –$\langle s - a_3\rangle$–$\langle a_3\rangle$.*

PROOF. By (10), after simple transformations we obtain $m_2 - m_1 = \beta + \gamma$, where

$$\beta = \binom{a_2}{s} - \binom{a_2+a_3-s}{s} + (a_0 - s)\binom{a_2+a_3-s}{s-1};$$

$$\gamma = \sum_{i=2}^{s-1}\left[\binom{a_0+a_1}{i} - \binom{a_1}{i} - \binom{s}{i}\right]\binom{a_2+a_3-s}{s-i}.$$

Since $a_0 \geq s-1$ and $a_3 \leq s-1$, $\beta \geq 0$. If $s=2$, then $\gamma = 0$ and $m_2 - m_1 \geq 0$. Suppose that $s \geq 3$. For $a_0 + a_1 > s$, either $a_0 = s-1$ and $a_1 \geq 2$ or $a_0 \geq s$; in both cases

$$\binom{a_0+a_1}{i} - \binom{a_1}{i} - \binom{s}{i} > 0$$

for $i \geq 2$; if in this case $a_2 + a_3 > s$, then $\gamma > 0$ and $m_2 - m_1 > 0$.

LEMMA 4. *If $a_0 \leq b_0$ and $a_2 \leq b_2$ (or $a_0 \leq b_2$ and $b_0 \geq a_2$), $a_1 \geq c$, $b_1 \geq c$, $a_0+c \geq s$, and $a_2+c \geq s$, then the total number of edges of the a.d.c. hypergraphs $\langle a_0\rangle$–$\langle a_1\rangle$–$\langle a_2\rangle$ and $\langle b_0\rangle$–$\langle b_1\rangle$–$\langle b_2\rangle$ is not greater than the total number of edges of the a.d.c. hypergraphs $\langle a_0\rangle$–$\langle c\rangle$–$\langle a_2\rangle$ and $\langle b_0\rangle$–$\langle b_1 + a_1 - c\rangle$–$\langle b_2\rangle$ [4].*

LEMMA 5. *For every d.c. $H(n,s,d)$ of the form of (1) with $d > 4$ there is a d.c. $H(n,s,d)$ of the form*

$$\langle 1\rangle–\langle b_1\rangle–\langle b_2\rangle–\cdots–\langle b_{d-1}\rangle–\langle 1\rangle, \tag{11}$$

for which the number of edges is not less than (greater than, if $s \geq 3$ and $a_i + a_{i+1} > s$ for $i = 1, \dots, d-2$) for a hypergraph of the form (1), and there exists r, $1 \leq r \leq d-2$, such that $b_r + b_{r+1} = s$.

PROOF. Consider the "initial part" $\langle 1\rangle$–$\langle a_1\rangle$–$\langle a_2\rangle$–$\langle a_3\rangle$–$\langle a_4\rangle$ of a hypergraph of the form (1), where $a_1 \geq s-1$.

If $a_4 \leq s-1$, then by Lemma 3 the number of edges in $\langle a_1\rangle$–$\langle a_2\rangle$–$\langle a_3\rangle$–$\langle a_4\rangle$ is not greater than in an a.d.c. $\langle a_1\rangle$–$\langle a_2+a_3+a_4-s\rangle$–$\langle s-a_4\rangle$–$\langle a_4\rangle$. It follows that (1) has no more edges than (11), where $b_i = a_i$ (for $i \neq 2,3$), $b_2 = a_2+a_3+a_4-s$, $b_3 = s - a_4$, and $r = 2$.

For $a_{d-4} \leq s-1$ the proof is similar (we do not distinguish between the a.d.c. hypergraphs $\langle a_0\rangle$–$\langle a_1\rangle$–$\cdots$–$\langle a_d\rangle$ and $\langle c_0\rangle$–$\langle c_1\rangle$–$\cdots$–$\langle c_d\rangle$, where $a_i = c_{d-i}$ for $i = 0, \dots, d$).

Suppose that $a_4, a_{d-4} > s-1$. If $a_2 \leq a_{d-2}$ (for $a_2 \geq a_{d-2}$ the proof is similar), then by Lemma 4 the total number of edges in $\langle 1\rangle$–$\langle a_1\rangle$–$\langle a_2\rangle$ and

$\langle a_{d-2}\rangle$–$\langle a_{d-1}\rangle$–$\langle 1\rangle$ is not greater than the total number of edges in the a.d.c. hypergraphs $\langle 1\rangle$–$\langle s-1\rangle$–$\langle a_2\rangle$ and $\langle a_{d-2}\rangle$–$\langle a_{d-1}+a_1-s+1\rangle$–$\langle 1\rangle$. Consider the a.d.c. $\langle 1\rangle$–$\langle s-1\rangle$–$\langle a_2\rangle$–$\langle a_3\rangle$–$\langle a_4\rangle$. By Lemma 1 it has no more edges than the a.d.c. $\langle 1\rangle$–$\langle s-1\rangle$–$\langle 1\rangle$–$\langle a_2+a_3-1\rangle$–$\langle a_4\rangle$. It follows that (1) has no more edges than (11), where $b_i = a_i$ ($i \neq 1, 2, 3$ or $d-1$), $b_1 = s-1$, $b_2 = 1$, $b_3 = a_2+a_3-1$, $b_{d-1} = a_{d-1}+a_1-s+1$, and $r = 1$.

For $s \geq 3$ and $a_i + a_{i+1} > s$ for $i = 1, \dots, d-2$ the proof is now obvious.

LEMMA 6. *If for $H(n,s,d)$ of the form of* (1)

$$a_r + a_{r+1} = s, \tag{12}$$

where $1 \leq r \leq d-2$, then

$$\alpha = \binom{a_{r-1}+a_r}{s} + 1 + \binom{a_{r+1}+a_{r+2}}{s} - \binom{a_{r-1}+a_{r+2}}{s} \leq 2,$$

where $\alpha = 2$ if and only if $a_r = a_{r+2}$ or $a_{r-1} = a_{r+1}$.

PROOF. From (9) and (12) it follows that $a_{r-1} = a_{r+1}+l$ and $a_{r+2} = a_r + p$, where $l, p \geq 0$ (note that $a_{r-1}+a_{r+2} \geq s$); then

$$\alpha - 2 = \binom{s+l}{s} + \binom{s+p}{s} - \binom{s+l+p}{s} - 1.$$

Clearly $\alpha - 2 = 0$ if $l = 0$ or $p = 0$, and $\alpha - 2 < 0$ if $l, p > 1$.

THEOREM 3. *If m is the number of edges in a hypergraph $H(n,s,d)$, then*

$$m \leq \begin{cases} \binom{n}{s} & \text{for } d = 1,\\ \binom{n-1}{s} + \binom{n-2}{s-1} & \text{for } d = 2,\\ d-2+2\binom{n-ks+s-1}{s} - \binom{n-ks}{s} & \text{for } d = 2k, k > 1,\\ d-2+\binom{n-ks+s-2}{s} + \binom{n-ks-1}{s-1} & \text{for } d = 2k+1, k > 0\ [4]. \end{cases}$$

The class of max.d.c. graphs (that is, uniform hypergraphs of rank 2) is described by Ore in [5].

THEOREM 4. *Suppose $s \geq 3$. Then*

a d.c. $H(n,s,d)$, where $d = 1, 2$, is maximal, and

a d.c. $H(n,s,d)$ of the form (1), *where $d \geq 3$, is maximal*

if and only if the following condition is satisfied:

if $d = 2k$ and $n - ks > s-1$ (case 1), then there exists t, $1 \leq t \leq k-1$, such that

$$a_i = \begin{cases} n - ks & \text{for } i = 2t,\\ 1 & \text{for } i \text{ even and } i \neq 2t,\\ s-1 & \text{for } i \text{ odd}, \end{cases} \tag{13}$$

where $i = 0, \dots, d$;

if $d = 2k$, $n - ks = s - p$ *and* $0 < p < s$ (*case* 2),
then there exist even numbers l *and* t, $2 \le l \le t \le d-2$, *such that*

$$a_i = \begin{cases} 1 & \text{for } i \text{ even and } (i < l \text{ or } i > t), \\ s-1 & \text{for } i \text{ odd and } (i < l \text{ or } i > t), \\ s-p & \text{for } i \text{ even and } l \le i \le t, \\ p & \text{for } i \text{ odd and } l < i < t, \end{cases}$$

where $i = 0, \dots, d$;
if $d = 2k+1$ (*case* 3), *then there exists* t, $1 \le t \le d-1$, *such that*

$$a_i = \begin{cases} n-ks-1 & \text{for } i = t, \\ 1 & \text{for } (i \text{ even and } i < t) \text{ or } (i \text{ odd and } i > t), \\ s-1 & \text{for } (i \text{ odd and } i < t) \text{ or } (i \text{ even and } i > t), \end{cases}$$

where $i = 0, \dots, d$.

The proof is obvious for $d = 1, 2$.

Case 1. The sufficiency is established by using formula (10) to count the number of edges in a d.c. $H(n, s, d)$ of the form (1) with the a_i as given in (13).

For $d = 4$ the necessity is easy to prove by Lemma 3. We suppose that the necessity holds for $d = 2(k-1)$ and prove it for $d = 2k$.

For a max.d.c. $H' = H(n, s, 2k)$ of the form (1), (12) must hold for some r, $1 \le r \le 2(k-1)$: otherwise by Lemma 5 there would exist a hypergraph $H(n, s, 2k)$ with a greater number of edges.

We consider a d.c. $H(n-s, s, 2(k-1))$ of the form

$$\langle 1\rangle - \langle a_1\rangle - \cdots - \langle a_{r-1}\rangle - \langle a_{r+2}\rangle - \cdots - \langle a_{2k-1}\rangle - \langle 1\rangle, \tag{14}$$

obtained from H' by deleting $\langle a_r\rangle - \langle a_{r+1}\rangle$. Since $a_{r-1} + a_{r+2} \ge s$ (see the proof of Lemma 6), such a hypergraph exists, and the number m_1 of its edges is related to the number m of edges in H' by the relation $m = m_1 + \alpha$, where α is the same as in Lemma 6.

Clearly m attains its maximum value if and only if 1) the hypergraph (14) is max.d.c., and 2) $\alpha = 2$ (if it is possible to satisfy conditions 1) and 2) simultaneously).

By Lemma 6, $\alpha = 2$ if and only if $a_r = a_{r+2}$ or $a_{r-1} = a_{r+1}$. By the induction hypothesis, since the hypergraph (14) is max.d.c., there exists t, $1 \le t \le k-1$, $2t \ne r, r+1$ such that

$$a_i = \begin{cases} n - s - (k-1)s = n - ks & \text{for } i = 2t, \\ 1 & \text{for } i \text{ even }, i \ne 2t, \\ s-1 & \text{for } i \text{ odd }, \end{cases} \tag{15}$$

where $i = 0, \dots, r-1, r+2, \dots, d$.

1. Suppose that $a_r = a_{r+2}$. Then $a_{r+2} \ne n - ks$: otherwise we would have $a_r = n-ks > s-1$ and $a_r + a_{r+1} > 1$, contradicting (12). It follows, by (15), that either $a_{r+2} = s-1$ or $a_{r+2} = 1$. In the first case $a_r = s-1$ and, according to (12), $a_{r+1} = 1$; and in this case, according to (15) either $a_{r-1} = n - ks$ or $a_{r-1} = 1$. In the second case, arguing similarly we obtain $a_r = 1$ and $a_{r+1} = a_{r-1} = s-1$.

2. Suppose that $a_{r-1} = a_{r+1}$. Then $a_{r-1} \neq n - ks$. If $a_{r+1} = s - 1$, then $a_r = 1$ and either $a_{r+2} = n - ks$ or $a_{r+2} = 1$. If $a_{r+1} = 1$, then $a_r = a_{r+2} = s - 1$.

3. Suppose $a_r = a_{r+2}$ and $a_{r-1} = a_{r+1}$. Then it follows from 1 and 2 that either $a_r = a_{r+2} = s - 1$ and $a_{r-1} = a_{r+1} = 1$ or $a_r = a_{r+2} = 1$ and $a_{r-1} = a_{r+1} = s - 1$.

In each of the cases 1–3, for a max.d.c. $H(n, s, 2k)$ of the form (1), the numbers a_i, $i = 0, \ldots, d$, are defined by (13).

In cases 2 and 3 the proof is similar.

REFERENCES

1. Claude Berge, *Graphes et hypergraphes*, Dunod, Paris, 1970; English transl., North-Holland, 1973.

2. A. A. Zykov, *Hypergraphs*, Uspekhi Mat. Nauk **29** (1974), no. 6(180), 89–154; English transl. in Russian Math. Surveys **29** (1974).

3. O. G. Rudenskaya, *The diameter of a homogeneous connected hypergraph*, Prikl. Mat. i Programmirovanie Vyp. 13 (1975), 139–144. (Russian)

4. ——, *Upper bound for the number of edges of a homogeneous connected hypergraph with a given diameter*, Kibernetika (Kiev) **1979**, no. 4, 146; English transl. in Cybernetics **15** (1979).

5. Oystein Ore, *Diameters in graphs*, J. Combinatorial Theory **5** (1968), 75–81.

Translated by T. R. WALSH

Amer. Math. Soc. Transl.
(2) Vol. **132**, 1986

On a Theorem from the Metric Theory of Continued Fractions

UDC 519.21

T. M. ZUPAROV

Let $[a_1, a_2, \ldots]$ be the continued fraction that represents $x \in (0,1]$. $a_n = a_n(x)$ can be regarded as a random variable on an appropriate probability space. Let $\{\varphi(n)\}$ be any sequence of positive numbers and $A(N,x)$ the number of integers $n \leq N$ that satisfy the inequality

$$a_n(x) \geq \varphi(n). \tag{1}$$

A classical theorem of Bernstein [1] states that $A(N,x) \to \infty$ as $N \to \infty$ with probability 1 if the series $\sum_{n=1}^{\infty} 1/\varphi(n)$ diverges and, conversely, for almost all x inequality (1) holds at most a finite number of times if $\sum_{n=1}^{\infty} 1/\varphi(n)$ converges.

We note that if the random variables a_n were independent, then Bernstein's theorem would follow from the Borel-Cantelli lemma since

$$0 < c_1 \leq \varphi(n)\lambda\{x; a_n(x) \geq \varphi(n)\} \leq c_2,$$

where λ is Lebesgue measure. Similarly, the law of the iterated logarithm and the central limit theorem would be applicable, since the characteristic functions of the set (1) are uniformly bounded.

Although the random variables $a_n(x)$ are not independent, as shown in [2] they satisfy certain conditions of weak dependence. Applying this property of $a_n(x)$, Walter Philipp obtained the law of the iterated logarithm and the central limit theorem (c.l.t.).

In this note an estimate of the remainder term in the c.l.t. for $A(N,x)$ is obtained.

1980 *Mathematics Subject Classification* (1985 *Revision*). Primary 11K50; Secondary 60F05.

Translation of Izv. Akad. Nauk UzSSR Ser. Fiz.-Mat. Nauk **1981**, no. 6, 9–12; MR **83h:**10096.

Let T be the map defined by the formula $Tx = x^{-1} \pmod 1$ for all $x \in (0,1)$. It follows from the definition that $a_2(x) = a_1(Tx)$, and so $a_{n+1}(x) = a_1(T^n x)$ for any integer $n \geq 1$. Here T^n is the n-fold iteration of the transformation T, defined by $T^n = T(T^{n-1})$. Consequently, T is the shift transformation

$$Tx = T(a_1(x), a_2(x), \dots] = [a_2(x), a_3(x), \dots].$$

The fundamental intervals of rank k

$$I_k = \{x; a_1(x) = r_1, a_2(x) = r_2, \dots, a_k(x) = r_k\}$$

generate the σ-algebra of the measurable sets.

We define a measure P on $(0,1)$ by setting

$$P(A) = \frac{1}{\ln 2} \int_A \frac{dx}{1+x}, \tag{2}$$

for any measurable set A. This is the well-known Gauss measure. As we know, P is equivalent to Lebesgue measure and is invariant under T.

We consider the probability space $(\Omega, \mathcal{B}, P)$, where Ω consists of the irrational numbers in $(0,1)$, $\mathcal{B}$ is the σ-algebra of Lebesgue-measurable sets, and the measure P is defined by (2).

We denote by $M_{a,b}$ the σ-algebra generated by the fundamental intervals I_k of rank k with $a \leq k \leq b$.

THEOREM 1. *Let $\varphi(n) \to \infty$ as $n \to \infty$ be a sequence of integers with $\sum_{n=1}^{\infty} 1/\varphi(n) = +\infty$. Set*

$$\Phi(N) = \frac{1}{\ln 2} \sum_{n=1}^{N} \ln\left(1 + \frac{1}{\varphi(n)}\right).$$

Then

$$P\left\{\frac{A(N,x) - \Phi(N)}{\sqrt{\Phi(N)}} < z\right\} = \frac{1}{\sqrt{2\pi}} \int_{-\infty}^{z} e^{-t^2/2}\, dt + O(\Phi^{-1/2}(N) \ln \Phi(N)).$$

To prove Theorem 1 we need the following modified variant of Theorem 3.1 in [3]:

THEOREM 2. *Let $\xi_{n1}, \dots, \xi_{nN}$ be a sequence of series of random variables with $\mathsf{E}\xi_{ni} = 0$, $i = 1, 2, \dots, N$, $\sup_i |\xi_{ni}| \leq 1$, and $B_N^2 = \mathsf{E}(\sum_{k=1}^{N} \xi_{nk})^2 \to \infty$ as $N \to \infty$. Denote by $M_{a,b}^{(N)}$ the σ-algebra generated by events of the form $\{\xi_{nk} < a\}$, $1 \leq a \leq k \leq b \leq N$. Assume that for any events $A \in M_{1,t}^{(N)}$ and $B \in M_{t+N,N}^{(N)}$*

1) *$|P(AB) - P(A)P(B)| \leq c_3 e^{-\lambda n} P(A)P(B)$, where $\lambda > 0$ is a constant.*

In addition, assume that

$$\sum_{k=l+1}^{l+H} \mathsf{E}|\xi_{kn}| = O\left(\mathsf{E}\left(\sum_{k=l+1}^{l+H} \xi_{nk}\right)^2\right), \qquad H \to \infty$$

uniformly in $l = 0, 1, 2, \ldots$. *Then*

$$P\left\{B_N^{-1}\sum_{k=1}^{N}\xi_{nk} < z\right\} = \frac{1}{\sqrt{2\pi}}\int_{-\infty}^{z} e^{-t^2/2}\,dt + O(B_N^{-1}\ln B_N).$$

Theorem 2 can be proved in exactly the same way as Theorem 3.1 in [3] by applying Charles Stein's theorem [4], where the term $\mathsf{E}|\mathsf{E}(G/\mathcal{E})|$ can be estimated by the following lemma:

LEMMA 1. *Let ξ, η, and ς be random variables measurable with respect to $M_{a,b}^{(N)}$, $M_{c,d}^{(N)}$, and $M_{e,k}^{(N)}$, respectively, where $c - b \geq m$ and $e - d \geq m$, and let $g\colon R \to R$ and $f\colon R^2 \to R$. Then*

$$|\mathsf{E}f(\xi,\varsigma)g(\eta) - \mathsf{E}f(\xi,\varsigma)\mathsf{E}g(\eta)| \leq c_4 e^{-\lambda m}\|f\|\mathsf{E}|g(\eta)| + c_4 e^{-\lambda m}\|g\|\mathsf{E}|f(\xi,\varsigma)|.$$

The proof of the lemma is carried out in [5].

We note that if the random variables ξ_{ni} do not satisfy the condition of uniform boundedness, then we cannot apply Lemma 1, and so in [3] it is required that the more restrictive condition (1′) hold in place of condition (1).

PROOF OF THEOREM 1. Let $E_n = \{x \in \Omega\colon a_n(x) \geq \varphi(n)\}$, let $\chi(n)$ be the characteristic function of the event E_n, i.e.,

$$\chi_n = \begin{cases} 1, & \text{for } a_n(x) \geq \varphi(n), \\ 0, & \text{for } a_n(x) < \varphi(n), \end{cases}$$

and let $\xi_n = \chi_n - p_n$, where $p_n = P(E_n)$. It is then easy to calculate that

$$\mathsf{E}\xi_n = 0, \qquad \sup_n |\xi_n| \leq 1,$$

$$\mathsf{E}|\xi_n| = 2p_n(1-p_n), \qquad \mathsf{E}\xi_n^2 = p_n(1-p_n),$$

and since P is an invariant measure with respect to T, we have

$$\begin{aligned} p_n = P(E_n) &= P\{x; a_1(T^{n-1}x) \geq \varphi(n)\} \\ &= P\{x; a_1(x) \geq \varphi(n)\} = \frac{1}{\ln 2}\ln\left(1 + \frac{1}{\varphi(n)}\right). \end{aligned} \tag{3}$$

It follows from a well-known lemma ([2], Lemma 2) that condition 1) of Theorem 2 holds.

LEMMA 2. *Let M_{ab} be the σ-algebra generated by fundamental intervals of rank k with $a \leq k \leq b$. Then for any $A \in M_{1,t}$ and $B \in M_{t+N,\infty}$*

$$|P(AB) - P(A)P(B)| \leq c_3 e^{-\lambda n} P(A)P(B),$$

where $\lambda > 0$.

Using this lemma, by (3) we have

$$\begin{aligned} B_N^2 = \mathsf{E}\left(\sum_{n=1}^{N}\xi_n\right)^2 &= \sum_{n=1}^{N}\mathsf{E}\xi_n^2 + \sum_{1\leq n\neq k\leq N}\mathsf{E}\xi_n\xi_k \\ &= \Phi(N) - \sum_{n=1}^{N}p_n^2 - 2\theta_1\sum_{1\leq k<n\leq N}p_k p_n e^{-\lambda(n-k)}, \qquad |\theta_1| \leq c_5. \end{aligned}$$

Hence, applying the inequality $2p_kp_n \le p_k^2 + p_n^2$, we obtain

$$B_N^2 - \Phi(N) = O\left(\sum_{n=1}^{N} p_n^2\right).$$

Consequently, by (3) and as $\varphi(n) \to \infty$ we find that

$$\sum_{n=1}^{N} p_n^2 = \frac{1}{\ln^2 2}\sum_{n=1}^{N} \ln^2\left(1 + \frac{1}{\varphi(n)}\right) = o(\Phi(N)),$$

and so $B_N^2 \to \infty$.

If $\Phi(l, H) = \sum_{l+1}^{l+H} p_n$ for any $l = 0, 1, \ldots$, then

$$\frac{1}{2}\sum_{n=l+1}^{l+H} \mathsf{E}|\xi_n| = \Phi(l, H) - \sum_{n=l+1}^{l+H} p_n^2,$$

$$\mathsf{E}\left(\sum_{n=l+1}^{l+H} \xi_n\right)^2 = \Phi(l, H) + 2\theta_2 \sum_{n=l+1}^{l+H} p_n^2,$$

where θ_2 does not depend on H and l. Hence, since $\Phi(l, H) \to \infty$ as $H \to \infty$ for any $l = 0, 1, 2, \ldots$ and

$$\sum_{n=l+1}^{l+N} p_n^2 = o(\Phi(l, H)),$$

we have

$$\sum_{n=l+1}^{l+H} \mathsf{E}|\xi_n| = O\left(\mathsf{E}\left(\sum_{n=l+1}^{l+H} \xi_n\right)^2\right), \qquad H \to \infty$$

uniformly in $l = 0, 1, 2, \ldots$.

Thus, all the conditions of Theorem 2 hold, and so Theorem 1 is an immediate consequence of Theorem 2.

REFERENCES

1. A. Ya. Khinchin [Khintchine], *Continued fractions*, 4th ed., "Nauka", Moscow, 1978; English transl. of 3rd ed., Noordhoff, 1963, and Univ. of Chicago Press, Chicago, Ill., 1964.

2. Walter Philipp, *Some metrical theorems in number theory*. II, Duke Math. J. **37** (1970), 447–458, 788.

3. T. M. Zuparov, *Moment inequalities and estimates of the remainder term in the central limit theorem for sequences of weakly dependent random variables*, Limit Theorems for Random Processes and Statistical Inference (S. Kh. Sirazhdinov, editor), "Fan", Tashkent, 1981, pp. 69–87. (Russian)

4. Charles Stein, *A bound for the error in the normal approximation to the distribution of a sum of dependent random variables*, Proc. Sixth Berkeley Sympos. Math. Statist. and Probab. (1970/71; L. Le Cam et al., editors), Vol. II: Probability Theory, Univ. of California Press, Berkeley, Calif., 1972, pp. 583–602.

5. Louis H. Y. Chen, *Poisson approximation for dependent trials*, Ann. Probab. **3** (1975), 534–545.

Translated by R. LENET

Amer. Math. Soc. Transl.
(2) Vol. **132**, 1986

Surfaces with $p_g = 3$ and $K^2 = 3$. Part I

VALENTIN V. ILIEV

Abstract. The author gives a complete description of the surfaces of type I with $p_g = 3$ and $K^2 = 3$ and having very ample canonical class. He proves that every surface of this type can be isomorphically embedded as a surface of degree 6 in the weighted 3-dimensional space of type $(1,1,1,2)$. With the aid of this presentation the results of Horikawa (1976) are deduced in a simpler and more natural way. He also completely describes surfaces of this type which are 3-fold cyclic Galois covers of $\mathbf{P}^2$, and proves that the moduli space is irreducible and has dimension 19.

Let S be a minimal surface with $p_g = 3$ and $K^2 = 3$. Horikawa [8] proved that the canonical system $|K|$ either does not have base points or has exactly one base point. With respect to this we speak of surfaces of type I or surfaces of type II. In the first part of this paper we shall study surfaces of type I.

§1.

Let σ be an automorphism of the space $\mathbf{P}^3$ given by the formula

$$\sigma(x_0 : x_1 : x_2 : x_3) = (x_0 : x_1 : x_2 : -x_3). \tag{1}$$

Consider the class of surfaces Y of degree 6 in $\mathbf{P}^3$ which satisfy the following conditions:

A) Y is irreducible and nonsingular.

B) Y is σ-invariant.

C) The curve $H_3 = Y \cap v(x_3)$ is irreducible and nonsingular.

These surfaces are of the form $Y = v(F)$, where F is an irreducible smooth form of degree 6,

$$F = f_6(x_0 : x_1 : x_2) + f_4(x_0 : x_1 : x_2)x_3^2 + f_2(x_0 : x_1 : x_2)x_3^4 + cx_3^6, \qquad c \neq 0. \tag{2}$$

1980 *Mathematics Subject Classification* (1985 *Revision*). Primary 14J10, 14J25.
Translation of Serdica **6** (1980), 352–362; MR **83i:** 14026a.

Moreover, because of C) the form f_6 is also irreducible and smooth. Standard considerations show that the polynomials F form a nonempty open subset in $\mathbf{P}^{49}$ (49 is the number of coefficients of the form F up to a constant multiple $\neq 0$). For example, the polynomial $F = x_0^6 + x_1^6 + x_2^6 + x_3^6$ belongs to this set. Automorphisms σ and $\sigma^2 = \mathrm{Id}$ form a group obviously isomorphic to $\mathbf{Z}_2$. By means of this isomorphism and (1) we define an action of $\mathbf{Z}_2$ on Y. The set of the fixed points coincides with the curve H_3. Let X denote the quotient $Y/\mathbf{Z}_2$.

THEOREM 1.1. *Every surface X obtained in this way is minimal, has geometric genus $p_g = 3$, $K^2 = 3$, and is of type* I. *Moreover, its canonical class is ample. The one-canonical map Φ_K makes X into a 3-fold cover of $\mathbf{P}^2$; and the two-canonical map is an isomorphic embedding.*

PROOF. Let $\Omega^2_{s,Y} = (\Omega^2_Y)^{\bigotimes S}$ be the sheaf of 2-forms of weight s, $s \geq 1$.

Step 1. *Determination of a basis of the* $\mathbf{C}$*-linear space* $H^0(Y, \Omega^2_{s,Y})$.

We follow [5], Chapter III, §5.4. Let $U = Y \cap \{x_0 \neq 0\}$, $y_i = x_i/x_0$, $i = 1, 2, 3$, and

$$G(y_1, y_2, y_3) = F(x_0 : x_1 : x_2 : x_3)/x_0^6 = F(1 : y_1 : y_2 : y_3).$$

Let U_i be the set of points of U where $\partial G/\partial y_i \neq 0$. Obviously, for example, on U_1 the local parameters are y_2 and y_3, so that

$$\Omega^2_{s,Y}[U_1] = \mathbf{C}[U_1]\omega_1^{\bigotimes S}, \qquad \text{where } \omega_1 = (\partial G/\partial y_1)^{-1}\, dy_2 \wedge dy_3.$$

Similarly,

$$\Omega^2_{s,Y}[U_2] = \mathbf{C}[U_2]\omega_2^{\bigotimes S}, \qquad \omega_2 = (\partial G/\partial y_2)^{-1}\, dy_1 \wedge dy_3,$$

and

$$\Omega^2_{s,Y}[U_3] = \mathbf{C}[U_3]\omega_3^{\bigotimes S}, \qquad \omega_3 = -(\partial G/\partial y_3)^{-1}\, dy_1 \wedge dy_2.$$

But the relation $\sum(\partial G/\partial y_i)\, dy_i = 0$ shows that $\omega_i = \omega_j$ on $U_i \cap U_j$, and therefore all the ω_i glue together into a nowhere vanishing differential form on U; it is clear that $\Omega^2_{s,Y}[U] = \mathbf{C}[U]\omega_3^{\bigotimes S}$. Now we shall investigate which 2-forms of weight s from $\Omega^2_{s,Y}[U]$ remain regular, for example, under the following change of coordinates:

$$z_1 = 1/y_1, \qquad z_2 = y_2/y_1, \qquad z_3 = y_3/y_1, \tag{3}$$

where z_1, z_2, z_3 are affine coordinates on $V = Y \cap \{x_1 \neq 0\}$. 2-forms from $\Omega^2_{s,Y}[U]$ can be written as $\omega = P(y_1, y_2, y_3)\omega_3^{\bigotimes S}$, where P is a polynomial. Calculations show that ω is regular on V if and only if $\deg P \leq 2s$. In particular, if $H(z_1, z_2, z_3) = G(1/z_1, z_2/z_1, z_3/z_1)$, then on V we have

$$\omega_3^{\bigotimes S} = (-1)^s z_1^{2s}[(\partial H/\partial z_3)^{-1}\, dz_1 \wedge dz_2]^{\bigotimes S}.$$

It is clear that $sK_Y = (\omega_3^{\bigotimes S}) = 2sH_0$, the complete linear system $|sK_Y|$ consists of divisors of the form $(\Phi/x_0^{2s}) + 2sH_0$, where Φ is a form of degree $2s$ and is cut out by the complete linear system $|2sL_0|$, and $L = \{x_0 = 0\}$. The canonical $\mathbf{C}$-linear homomorphism $L(2sL_0) \to L(2sH_0) \to 0$ has kernel $J_s = M_Y \cap L(2sL_0)$, where $M_Y = (G) \subset \mathbf{C}[y_1, y_2, y_3]$ is the ideal of the affine surface U. It is easy to see that every polynomial $P \in L(2sL_0)$ is of the form $P = \sum a_{klt} y_1^k y_2^l y_3^t + QG$, where $0 \leq t \leq 5$, and $a_{klt} \in \mathbf{C}$ and $Q \in \mathbf{C}[y_1, y_2, y_3]$ are determined uniquely. If E_s denotes the subspace in $L(2sL_0)$ generated by the monomials of the form $y_1^k y_2^l y_3^t$, $0 \leq k+l+t \leq 2s$, $0 \leq t \leq 5$, then $L(2sL_0) = E_s \oplus J_s$ and $E_s \cong L(2sH_0)$. It is now obvious that

$$H^0(Y, \Omega^2_{s,Y}) = \bigoplus_{\substack{0 \leq k+l+t \leq 2s \\ 0 \leq t \leq 5}} \mathbf{C} y_1^k y_2^l y_3^t \omega_3^{\otimes S}$$

for $s \geq 1$.

Step 2. Determination of a basis of the $\mathbf{C}$*-linear space* $H^0(X, \Omega^2_{s,X})$.

LEMMA 1.1. $H^0(X, \Omega^2_{s,X}) \cong \{\sigma$*-invariant forms* $\omega \in H^0(Y, \Omega^2_{s,Y})$ *whose divisor is divisible by* $sH_3\}$.

PROOF. Let $\eta: Y \to X$ be the natural projection. The map

$$\eta^*: H^0(X, \Omega^2_{s,X}) \to H^0(Y, \Omega^2_{s,Y})$$

is an injective $\mathbf{C}$-linear homomorphism and the equality $(\eta^*\omega) = \eta^*(\omega) + sH_3$ shows that the image $\operatorname{Im} \eta^*$ is contained in the subspace defined on the right-hand side. We shall prove the other inclusion. Let $\omega \in H^0(Y, \Omega^2_{s,Y})$ be a σ-invariant 2-form of weight s, and let sH_3 divide the divisor (ω). If $\omega = P\omega_3^{\otimes S}$ then $P(y_1, y_2, y_3) = y_3^s \tilde{P}(y_1, y_2, y_3)$, where $\deg \tilde{P} \leq s$ (y_3 is a local equation of the curve H_3 in U). Since ω is σ-invariant, we have

$$P(y_1, y_2, y_3) = \tilde{P}_1(y_1, y_2, y_3^2)$$

and

$$\omega = \tilde{P}_1(y_1, y_2, y_3^2)[(\partial G/\partial y_2)^{-1}\, dy_1 \wedge dy_3^2],$$

i.e. $\omega \in \operatorname{Im} \eta^*$. The lemma is proved.

Furthermore, the equation $G = 0$ implies that the polynomial P has the representation

$$P = y_3^s \sum a_{klt} y_1^k y_2^l y_3^{2t}, \qquad 0 \leq k + l + 2t \leq s, \quad 0 \leq t \leq 2,$$

and this is unique because y_3 and G are relatively prime polynomials. Let E_s^* denote the subspace of E_s generated by the monomials of the form $y_1^k y_2^l y_3^{2t+s}$, $0 \leq k+l+2t \leq s$, $0 \leq t \leq 2$. For any $s \geq 1$ we have $y_3^s \in E_s^*$, and if $\omega = y_3^s \omega_3^{\otimes S}$ then the coordinate change (3) shows that

$$\omega = z_1^s z_3^s [(\partial H/\partial z_3)^{-1}\, dz_1 \wedge dz_2]^{\otimes S}$$

on V, i.e. $(\omega) = sH_0 + sH_3$. Near a point $x \in H_3$ we have

$$\omega = (-1)^s z_1^s z_3^s [(\partial H/\partial z_2)^{-1}\, dz_1 \wedge dz_3]^{\otimes S},$$

and since z_1, z_3^2 are local coordinates in a neighborhood of the point $y = \eta(x)$ the presentation

$$\omega = (-1/2)^s z_1^s [(\partial H/\partial z_2)^{-1}\, dz_1 \wedge dz_3^2]^{\otimes S}$$

shows that $(\omega)|_X = sH_0^*$, where $H_0^* = \eta(H_0)$. In particular, we can set $K_X = H_0^*$, and then

$$L(sK_X) = \bigoplus_{\substack{0\le k+l+2t\le s\\ 0\le t\le 2}} \mathbf{C}y_1^k y_2^l y_3^{2t};$$

the multiplication by $1/y_3^s$ gives an isomorphism $E_s^* \to L(sK_X)$. Moreover it is clear that

$$H^0(X, \Omega_{s,X}^2) = \bigoplus_{\substack{0\le k+l+2t\le s\\ 0\le t\le 2}} \mathbf{C}y_1^k y_2^l y_3^{2t+s}\omega_3^{\otimes S} \quad \text{for } s\ge 1.$$

Step 3. $p_g = 3$ *and* $K^2 = 3$.

PROOF. We already proved in Step 2 that $p_g = \dim_{\mathbf{C}} H^0(X, \Omega_X^2) = 3$. Assumption C) for the surface Y shows that H_3 is a smooth plane curve of degree 6. Then $g =$ the genus of $H_3 = 10$. On the other hand $K_Y \sim 2H_0 \sim 2H_3$ and therefore $g = \frac{1}{2}(H_3, H_3 + K_Y) + 1 = \frac{3}{2}H_3^2 + 1$, i.e. $H_3^2 = 6 = H_0^2$. Now $K_X = H_0^{*2} = \frac{1}{2}H_0^2$ implies $K_X^2 = 3$.

Step 4. *X is a minimal surface.*

PROOF. Suppose that X is not minimal. Then the Castelnuovo criterion ([4], Chapter II, §2) gives us a smooth irreducible rational curve Λ with $\Lambda^2 = -1$. Because $g(\Lambda) = 0$, the formula $g(\Lambda) = \frac{1}{2}(\Lambda, \Lambda + K_X) + 1$ gives $(\Lambda, H_0^*) = -1$. Therefore $(\eta^*\Lambda, H_0) = -2$, and since $\eta^*\Lambda$ is an effective divisor we have a contradiction (one can always find an irreducible hyperplane section of the surface Y which is not a component of the divisor $\eta^*\Lambda$).

Consideration in Step 2 shows that

$$L(K_X) = \mathbf{C} \oplus \mathbf{C}y_1 \oplus \mathbf{C}y_2,$$
$$L(2K_X) = \mathbf{C} \oplus \mathbf{C}y_1 \oplus \mathbf{C}y_2 \oplus \mathbf{C}y_1^2 \oplus \mathbf{C}y_1y_2 \oplus \mathbf{C}y_2^2 \oplus \mathbf{C}y_3^2,$$

and this choice of bases defines the canonical maps Φ_{K_X} and Φ_{2K_X}.

Step 5. *Φ_{K_X} is a* 3*-fold cover and the diagram*

$$\begin{array}{ccc} Y & \xrightarrow{j} & \mathbf{P}^3 \\ \eta\downarrow & & \downarrow \mathrm{pr} \\ X & \xrightarrow{\Phi_{K_X}} & \mathbf{P}^2 \end{array}$$

is commutaive (here $j{:}\,Y \to \mathbf{P}^3$ is the natural embedding).

PROOF. Obvious. This also proves

Step 6. *X is a surface of type* I.

Step 7. *$\Phi_{2K_X}{:}\,X \to \mathbf{P}^6$ is an isomorphic embedding.*

PROOF. Φ_{2K_X} is given by the formula

$$\Phi_{2K_X}(y) = \begin{cases} (1 : y_1(y) : y_2(y) : y_1^2(y) : y_1(y)y_2(y) : y_2^2(y) : y_3^2(y)); & y\in U^*,\\ (z_1^2(y) : z_1(y) : z_1(y)z_2(y) : 1 : z_2(y) : z_2^2(y) : z_3^2(y)); & y \in V^* \end{cases}$$

and therefore is a regular map. It is clear that on Y it is two-to-one set-theoretically and moreover is $\mathbf{Z}_2$-invariant, i.e., on X, Φ_{2K_X} is an injective map.

Calculating the Jacobian, we see that Φ_{2K_X} is a closed immersion; the proof is finished.

Step 7 shows that the divisor $2K_X$ is very ample, i.e. we have also proved

Step 8. *K_X is an ample divisor.*

The proof of Theorem 1.1 is finished.

§2. A converse of Theorem 1.1

Let $b_0, \ldots, b_n$ be natural numbers and G the subgroup of $PGL(n, \mathbf{C})$ consisting of matrices of the form

$$\begin{pmatrix} \exp(2\pi i\beta_0/b_0) & & \mathbf{0} \\ & \ddots & \\ \mathbf{0} & & \exp(2\pi i\beta_n/b_n) \end{pmatrix},$$

where $0 \leq \beta_i < b_i$ and $\beta_i \in \mathbf{Z}$ for $i = 0, 1, \ldots, n$.

DEFINITION 2.1 (see [11]). The quotient $\mathbf{P}^n/G$ is called the *n-dimensional weighted projective space of type* $(b_0, \ldots, b_n)$ and is denoted by $\mathbf{P}^n(b_0, \ldots, b_n)$.

DEFINITION 2.2. $f(w_0, \ldots, w_n)$ is called a *quasihomogeneous polynomial of type* $(b_0, \ldots, b_n)$ *and degree* d if it is a $\mathbf{C}$-linear combination of monomials of the form $w_0^{s_0} \cdots w_n^{s_n}$ with $s_0 b_0 + \cdots + s_n b_n = d$.

If $b_0 = 1$, $b_1 = 1$, $b_2 = 1$, and $b_3 = 2$, then $G \cong \mathbf{Z}_2$ and consists of automorphisms σ and $\sigma^2 = \mathrm{id}$ of the space $\mathbf{P}^3$ (see §1). It is now clear that the quotient $Y/\mathbf{Z}_2 = X$ is isomorphically embedded in $\mathbf{P}^3(1,1,1,2)$ so that the diagram

$$\begin{array}{ccc} Y & \xrightarrow[j]{} & \mathbf{P}^3 \\ \downarrow & & \downarrow \mathrm{pr} \\ X & \xrightarrow{j^\sigma} & \mathbf{P}^3(1,1,1,2) \end{array}$$

is commutative. Moreover if we set $w_0 = x_0$, $w_1 = x_1$, $w_2 = x_2$, and $w_3 = x_3^2$, and notice that $F(x_0 : x_1 : x_2 : x_3) = F^*(x_0, x_1, x_2, x_3^2)$, then $j\sigma(X) = v(F^*)$ in $\mathbf{P}^3(1,1,1,2)$, where $F^* = F^*(w_0, w_1, w_2, w_3)$ is a quasihomogeneous polynomial of type $(1,1,1,2)$ and degree 6. Therefore we can say that every surface in $\mathbf{P}^3(1,1,1,2)$ defined as the set of zeros of an irreducible smooth quasihomogeneous polynomial of type $(1,1,1,2)$ and degree 6 has invariants $p_g = 3$ and $K^2 = 3$, is of type I, and has the properties listed in Theorem 1.1. The main theorem of this paper is converse to Theorem 1.1.

THEOREM 2.1. *Let S be a minimal surface with $p_g = 3$ and $K^2 = 3$ of type* I *with ample canonical divisor K. Then S isomorphically embeds in* $\mathbf{P}^3(1,1,1,2)$ *as the set X consisting of all zeros of an irreducible smooth quasihomogeneous polynomial of type* $(1,1,1,2)$ *and degree* 6.

PROOF. We begin by stating two results of Horikawa ([8], §2).

LEMMA 2.1. *Φ_K is a 3-fold cover of* $\mathbf{P}^2$.

LEMMA 2.2. *Φ_{2K} is a regular birational isomorphism.*

Let $H^0(S, O(K)) = \mathbf{C}\lambda_0 \oplus \mathbf{C}\lambda_1 \oplus \mathbf{C}\lambda_2$. In this case the Kodaira formula for plurigenera ([6], §V) is $P_m = \frac{3}{2}m(m-1)+4$, $m \geq 2$, and therefore $P_2 = 7$. The sections $\lambda_i\lambda_j$, $0 \leq i \leq j \leq 2$, are linearly independent elements of the space $H^0(S, O(2K))$, which has dimension 7, and we can complete this system to obtain a basis

$$H^0(S, O(2K)) = \mathbf{C}\lambda_0^2 \oplus \mathbf{C}\lambda_1^2 \oplus \mathbf{C}\lambda_2^2 \oplus \mathbf{C}\lambda_0\lambda_1 \oplus \mathbf{C}\lambda_0\lambda_2 \oplus \mathbf{C}\lambda_1\lambda_2 \oplus \mathbf{C}\mu.$$

For our purposes we shall choose a global section μ of the sheaf $O(2K)$ in such a way that its divisor $C = (\mu)$ is a smooth irreducible curve. But Lemma 2.2 implies, in particular, that the image $\Phi_{2K}(S)$ is a surface; so the first Bertini theorem ([4], Chapter I, §3) guarantees the existence of such a section μ. Let $\{U_i\}$ be an open (in the complex-analytic topology) cover of the surface S such that the canonical bundle $\pi: K \to S$ is trivial on each U_i and has the transition functions $J_{ij} \in O^*(U_i \cap U_j)$. Suppose that the global section μ of the sheaf $O(2K)$ is given by functions $\mu_i \in O(U_i)$, $\mu_i = J_{ij}^2\mu_j$ in $U_i \cap U_j$, and μ_i is a local equation of the curve C on U_i. Moreover we can assume (after refining the cover $\{U_i\}$) that μ_i is one of the local coordinates u_i, v_i, say $\mu_i = u_i$ in those U_i for which $U_i \cap C \neq \varnothing$ (C is a smooth curve). Let T denote the compact surface which has equation $\xi_i^2 = \mu_i$ on $\pi^{-1}(U_i) \cong U_i \times \mathbf{C}$, where ξ_i is a fiber coordinate on the neighborhood U_i. Obviously $\pi: T \to S$ is a 2-fold cyclic cover with the ramification curve Γ, $\Gamma|_{\pi^{-1}(U_i)} = \{\xi_i = 0\}$, and a theorem of Grauert's guarantees that T is an algebraic surface ([12], §1). In $\pi^{-1}(U_i) \cap \pi^{-1}(U_j)$ we have $\xi_i = J_{ij}\xi_j$; therefore the functions $\xi_i \in O(\pi^{-1}(U_i))$ define a global section ξ of the induced bundle π^*K on T and $(\xi) = \Gamma$, i.e. $\Gamma \in \pi^*K$. Moroever, by construction, $\xi^2 = \mu$.

LEMMA 2.3. $H^0(T, O(m\pi^*K)) \simeq H^0(S, O((m-1)K)) \oplus H^0(S, O(mK))$, $m \geq 1$.

PROOF. Let σ be the involution of the bundle K defined by $\xi_i \to -\xi_i$. Obviously σ is compatible with the canonical projection $\pi: K \to S$ and the surface T is σ-invariant, i.e. σ is an involution of T. We have $\mathrm{Gal}(T/S) \cong (\sigma) \cong \mathbf{Z}_2$; the group $\mathbf{Z}_2$ acts on the $\mathbf{C}$-linear space $H^0(T, O(m\pi^*K))$ and splits it into the invariant and anti-invariant subspaces

$$H^0(T, O(m\pi^*K)) = H^0(T, O(m\pi^*K))^+ \oplus H^0(T, O(m\pi^*K))^-.$$

Obviously

$$H^0(T, O(m\pi^*K))^+ \cong \pi^*H^0(S, O(mk)).$$

To every section α' from $H^0(T, O((m-1)\pi^*K))$ we associate the section $\alpha' \otimes \xi$ from $H^0(T, O(m\pi^*K))$. Since ξ is a global section of π^*K not identically equal to zero on any open subset of T, we obtain an injective $\mathbf{C}$-linear homomorphism

$$H^0(T, O((m-1)\pi^*K) \to H^0(T, O(m\pi^*K)).$$

Clearly the image of the subspace $H^0(T, O((m-1)\pi^*K))^+$ is contained in $H^0(T, O(m\pi^*K))^-$. Now let $\alpha \in H^0(T, O(m\pi^*K))^-$, i.e. $\alpha^\sigma = -\alpha$, and let

$x \in \Gamma$. Then $x \in \pi^{-1}(U_i)$ with $U_i \cap C \neq 0$, and in $\pi^{-1}(U_i)$ the surface T has equation $\xi_i^2 = u_i$, where u_i, v_i is the local coordinate system chosen before. Suppose that this system is centered at the point $y = \pi(x)$, i.e. $v_i(y) = 0$. It is clear that ξ_i, v_i is a local coordinate system at point x and if $\alpha = \alpha_i(\xi_i, v_i)$ then $\alpha^\sigma = \alpha_i(-\xi_i, v_i)$ near x. Consider the Taylor expansion for the function α_i with respect to ξ_i, v_i:

$$\alpha_i(\xi_i, v_i) = \sum a_{kl}\xi_i^k v_i^l.$$

Then

$$\alpha_i^\sigma = \alpha_i(-\xi_i, v_i) = \sum(-1)^k a_{kl}\xi_i^k v_i^l$$

and the equality $\alpha_i^\sigma = -\alpha_i$ shows that $a_{kl} = 0$ for $k \equiv 0 \pmod 2$, i.e., in a neighborhood of the point x we have $\alpha_i = \xi_i \alpha_i'$, where $\alpha_i' \in O_x$. Near a point $x \notin \Gamma$ we define α_i' by this equality (ξ_i is an invertible element of the ring O_x). Since the local rings O_x are factorial, the equality

$$\alpha_i'\xi_i = J_{ij}^m \alpha_j' \xi_j = J_{ij}^{m-1}\alpha_j' J_{ij}\xi_j = J_{ij}^{m-1}\alpha_j'\xi_i$$

in $\pi^{-1}(U_i) \cap \pi^{-1}(U_j)$ implies $\alpha_i' = J_{ij}^{m-1}\alpha_j'$, i.e., the functions $\alpha_i' \in O(\pi^{-1}(U_i))$ define a global section α' from $H^0(T, O((m-1)\pi^*K))^+$ and, obviously, $\alpha = \alpha' \otimes \xi$. Finally we obtain isomorphisms

$$H^0(T, O(m\pi^*K))^- \cong H^0(T, O((m-1)\pi^*K))^+ \cong H^0(S, O((m-1)K)),$$

which implies the lemma.

COROLLARY 1. $\dim_{\mathbf{C}} H^0(T, O(m\pi^*K)) = P_{m-1}(S) + P_m(S)$ $(P_0 = 1)$.

COROLLARY 2. $H^0(T, O(\pi^*K)) = \mathbf{C}\lambda_0 \oplus \mathbf{C}\lambda_1 \oplus \mathbf{C}\lambda_2 \oplus \mathbf{C}\xi$.

Therefore the diagram

$$\begin{array}{ccc} T & \xrightarrow{\Phi_{\pi^*K}} & \mathbf{P}^3 \\ \pi\downarrow & & \downarrow \mathrm{pr} \\ S & \xrightarrow{\Phi_K} & \mathbf{P}^2 \end{array} \tag{4}$$

is commutative. This makes the next lemma obvious.

LEMMA 2.4. *Φ_{π^*K} is a regular map, and the image $Y = \Phi_{\pi^*K}(T)$ is an irreducible closed σ-invariant surface in $\mathbf{P}^3$.*

LEMMA 2.5. *The surface T is minimal and has invariants $p_g = 10$ and $K_T^2 = 24$.*

PROOF. Let Λ be an exceptional curve of the first kind on T, i.e. $g(\Lambda) = 0$ and $\Lambda^2 = -1$. The adjunction formula gives $(\Lambda, K_T) = -1$. On the other hand, $K_T \sim 2\pi^*K$ and therefore $2(\Lambda, \pi^*K) = -1$, a contradiction. Now Lemma 2.3 implies

$$p_g = \dim_{\mathbf{C}} H^0(T, O(2\pi^*K)) = 10 \quad \text{and} \quad K_T^2 = 4(\pi^*K)^2 = 8K^2 = 24.$$

LEMMA 2.6. *If the divisor K is ample, so is K_T.*

The proof is a simple application of the Nakai-Moishezon criterion ([7], Chapter V). It is also easy to check the following fact.

LEMMA 2.7. *Φ_{2K} is injective in an open set $U \subset S$ if and only if Φ_{π^*K} is injective in $\pi^{-1}(U) \subset T$.*

LEMMA 2.8. *If $\pi(x) = y$ then Φ_{2K} is an immersion at the point y if and only if Φ_{π^*K} is an immersion at x.*

PROOF. Compare the ranks of the matrices of the cotangent maps corresponding to the maps Φ_{2K} and Φ_{π^*K}.

Lemma 2.2 implies that there exist open sets $U \subset S$ and $V \subset S'$, where $S' = \Phi_{2K}(S) \subset \mathbf{P}^6$, such that $\Phi_{2K}: U \to V$ is an isomorphism. Let $H_i(z_0 : \cdots : z_6) = 0$, $i = 1, \ldots, s$, be equations of the set $S' - V$ in $\mathbf{P}^6$. Then $H_i(\lambda_0^2 : \cdots : \xi^2) = 0$, $i = 1, \ldots, s$, are equations of the set $S - U$ in $T - \pi^{-1}(U)$. Let W denote the complement of the subset defined in Y by these equations. Then Lemma 2.7 implies that $\Phi_{\pi^*K}: \pi^{-1}(U) \to W$ is a regular bijection. On the other hand, diagram (4) defines the dual diagram of the fields, and Lemma 2.4 gives

$$[\mathbf{C}(Y) : \mathbf{C}(\mathbf{P}^2)] = 2 \text{ or } 6.$$

Suppose first that $[\mathbf{C}(Y) : \mathbf{C}(\mathbf{P}^2)] = 2$. Let $z \notin \mathrm{pr}(Y - W) \cup \Phi_K(C) \cup \{$the branch locus of $\Phi_K\}$. Then $\#\pi^{-1}(\Phi_K^{-1}(z)) = 6$ and $\#\Phi_{\pi^*K}^{-1}(\mathrm{pr}^{-1}(z)) \leq 2$, a contradiction. Therefore $[\mathbf{C}(Y) : \mathbf{C}(\mathbf{P}^2)] = 6$, $[\mathbf{C}(T) : \mathbf{C}(Y)] = 1$, and we have obtained

LEMMA 2.9. *Φ_{π^*K} is a regular birational isomorphism, and its image Y is a surface of degree 6 in $\mathbf{P}^3$.*

Let $Y = V(F)$, $F = F(x_0 : x_1 : x_2 : x_3)$, be a form of degree 6 written as (2). Then

$$Y \cong \mathrm{Proj}(\mathbf{C}[x_0, x_1, x_2, x_3]/(F)).$$

Let R_m denote $H^0(T, O(m\pi^*K))$, and let $R = \bigoplus_{m \geq 0} R_m$ be the corresponding graded ring. Note that $R_0 = \mathbf{C}$ and $R_1 = \mathbf{C}\lambda_0 \oplus \mathbf{C}\lambda_1 \oplus \mathbf{C}\lambda_2 \oplus \mathbf{C}\xi$.

LEMMA 2.10. *The $\mathbf{C}$-algebra R is generated by homogeneous elements of degree one: $R = \mathbf{C}[\lambda_0, \lambda_1, \lambda_2, \xi]$.*

PROOF. Obviously $F(\lambda_0, \lambda_1, \lambda_2, \xi) = 0$ is the only algebraic relation between $\lambda_0, \lambda_1, \lambda_2$, and ξ; this implies that $\dim_{\mathbf{C}} R_1^m = \binom{m+3}{3} - \binom{m-3}{3}$ (we set $\binom{m-3}{3} = 0$ for $m < 6$). But Lemma 2.3 and Corollary 1 give

$$\dim_{\mathbf{C}} R_m = P_m + P_{m-1} = \tfrac{3}{2}m(m-1) + 4 + \tfrac{3}{2}(m-1)(m-2) + 4 = 3(m-1)^2 + 8$$

for $m \geq 3$, and $\dim_{\mathbf{C}} R_2 = 10$; therefore $\dim_{\mathbf{C}} R_1^m = \dim_{\mathbf{C}} R_m$ for $m \geq 1$. The lemma is proved.

It is now clear that $\mathbf{C}[x_0, x_1, x_2, x_3]/(F) \cong R$ and

$$\mathrm{Proj}(\mathbf{C}[x_0, x_1, x_2, x_3]/(F)) \cong \mathrm{Proj}\, R.$$

Furthermore,

$$\operatorname{Proj} R = \operatorname{Proj}\Big(\bigoplus_{m\geq 0} R_m\Big) \cong \operatorname{Proj}\Big(\bigoplus_{m\geq 0} R_{2m}\Big)$$
$$\cong \mathbf{Z} = \operatorname{Proj}\Big(\bigoplus_{m\geq 0} H^0(T, O(mK_T))\Big).$$

The last scheme is an abstract canonical model of the surface T (see [6], §I), and we see that the image Y is isomorphic to the scheme Z. Let $E_1, \dots, E_t$ denote all the curves with $K_T \cdot E_i = 0$. The set $\{E_i\}$ is characterized by two conditions ([6], §II):

a) E_i is a rational curve.

b) $E_i^2 = -2$.

M. Artin proved (see [6], §II) that there exists a regular map $\delta: T \to Z$ with the following properties:

c) The restriction $\delta: T - U_1^t E_i \to Z - \operatorname{Sing}(Z)$ is an isomorphism.

d) If $z \in \operatorname{Sing}(Z)$ then its preimage $\delta^{-1}(z)$ is equal to $\bigcup E_i$, where $\bigcup E_i$ is a connected component of the union $\bigcup_1^t E_i$. Moreover the set $\operatorname{Sing}(Z)$ of the singular points consists of rational double singularities. Since the surface T is minimal, the map $\delta: T \to Z$ is a minimal resolution of singularities of Z. The commutativity of the diagram

$$\begin{array}{ccc} T & \xrightarrow{\Phi_{\pi^*K}} & Y \\ \delta\downarrow & \nearrow_{\cong} & \\ Z & & \end{array}$$

proves

LEMMA 2.11. *Y is a surface of degree 6 in $\mathbf{P}^3$ with rational double points, and the map $\Phi_{\pi^*K}: T \to Y$ is the minimal resolution of its singularities.*

In our case the divisor K_T is ample, i.e. the set $\{E_i\}$ is empty, and therefore the map $\Phi_{\pi^*K}: T \to Y$ is an isomorphism. In particular, Y is a smooth surface. Since $\Phi_{\pi^*K} \circ \sigma = \sigma \circ \Phi_{\pi^*K}$, the surface Y is σ-invariant and therefore the quotient map $\Phi^{\sigma}_{\pi^*K}: S \to X$ is an isomorphism (here $X = Y/\mathbf{Z}_2$). Theorem 2.1 is proved.

Theorems 2.1 and 1.1 immediately imply

COROLLARY 1. *Under the assumption of Theorem 2.1, Φ_{2K} is an isomorphic embedding.*

This is a refinement of Horikawa's result (Lemma 2.2).

§3. A description of surfaces with $p_g = 3$ and $K^2 = 3$ for which the one-canonical map Φ_K is a 3-fold cyclic Galois cover of $\mathbf{P}^2$ (for short, a CGC-surface)

Let S be a CGC-surface. Then by definition there exists a smooth curve B such that the restriction of the map $\pi = \Phi_K$ to $S - B$ is an unramified 3-fold cover of $\mathbf{P}^2 - \Lambda$, $\Lambda = \pi(B)$, and $\pi|_B: B \to \Lambda$ is an isomorphism. B is the ramification curve for π, and Λ is the branch locus.

LEMMA 3.1. *Λ is a curve of degree 6 in $\mathbf{P}^2$.*

PROOF. Suppose a curve l in $\mathbf{P}^2$ intersects Λ transversally and $\pi^* l = K$ is an irreducible nonsingular curve from the canonical linear system $|K|$ (the first Bertini theorem guarantees the existence of such a curve l). Then $\pi|_K: K \to l$ is a 3-fold Galois cover ramified at points of $K \cap B$ which have ramification index 2. Obviously $\# K \cap B = \# l \cap \Lambda$. The Hurwitz formula (see [7], Chapter IV, §2) gives

$$2g(K) - 2 = 3[2g(l) - 2] + \sum_{x \in l \cap \Lambda} 2.$$

On the other hand we have $g(l) = 0$ and $g(K) = \frac{1}{2}(K, K + K) + 1 = K^2 + 1 = 4$, i.e. $\# l \cap \Lambda = 6$ and therefore $\deg \Lambda = 6$.

By a result of Wavrik ([12], Theorem 1.2) there exist a line bundle N on $\mathbf{P}^2$ and an open (in the complex-analytic topology of $\mathbf{P}^2$) cover $\{V_j\}$ such that N trivializes on each V_j, and the surface S can be identified with a subvariety of the bundle N given by the equations $\eta_j^3 = \Phi_j$, where η_j is a fiber coordinate and Φ_j is the local equation of the curve Λ in V_j.

Since the group $H^1(\mathbf{P}^2, O^*_{\mathbf{P}^2})$ of the classes of isomorphic line bundles is isomorphic to the group $\mathbf{Z}$ of integers and $N^3 = O_{\mathbf{P}^2}(6)$, we can assume that $N = O_{\mathbf{P}^2}(2)$. The line bundle $\pi: O_{\mathbf{P}^2}(2) \to \mathbf{P}^2$ trivializes on the standard cover $\mathbf{P}^2 = U_0 \cup U_1 \cup U_2$, where $U_i = \{x_i \neq 0\}$, and in $U_i \cap U_k$ it has the transition functions $s_{ik} = (x_k/x_i)^2$. It follows from the proof of Wavrik's theorem that we can assume the cover $\{V_j\}$ refines the cover $\{U_i\}$. Because $H^1(\mathbf{P}^2, O^*_{\mathbf{P}^2}) = H^1(\{U_i\}, O^*_{\mathbf{P}^2})$, it is easy to see that the following holds:

PROPOSITION 3.1. *If S is a CGC-surface then it maps isomorphically onto the surface in $O_{\mathbf{P}^2}(2)$ given in the standard trivialization $\pi^{-1}(U_i) \cong U_i \times \mathbf{C}$, $U_i = \{x_i \neq 0\}$, $i = 0, 1, 2$, by the equation $\mu_i^3 = G_i$, where μ_i is a fiber coordinate and $G_i = G/x_i^6$ is a smooth form of degree 6.*

Now we shall prove the converse.

PROPOSITION 3.2. *Let S be a surface in $O_{\mathbf{P}^2}(2)$ given in the standard trivialization by the equation $\mu_i^3 = G_i$, $G_i = G/x_i^6$, where G is a smooth form of degree 6. Then S is a minimal surface with invariants $p_g = 3$ and $K^2 = 3$, and S is a CGC-surface.*

PROOF. Let σ denote the automorphism of the bundle $O_{\mathbf{P}^2}(2)$ given by the formula $\mu_i \to \varepsilon \mu_i$, where ε is a primitive third root of unity. It is clear that the surface S is σ-invariant and that σ induces an automorphism of this surface (which we also denote σ). The cyclic group of order three (σ) is isomorphic to $\mathbf{Z}_3$ and obviously is the Galois group of the cyclic Galois cover $S \xrightarrow{\pi} \mathbf{P}^2$. The ramification curve $B = S \cap \{\mu_i = 0\}$ of this cover is isomorphic to the curve

$\Lambda = \{G = 0\} \subset \mathbf{P}^2$. Moreover we have $K_S \sim \pi^*(K_P) + 2B \sim \pi^*(-3l_0) + 2B = -3\pi^* l_0 + 2B$, where $l_0 = \{x_0 = 0\} \subset \mathbf{P}^2$. The function μ_0 is rational on S (it is regular on $\pi^{-1}(U_0)$), and the equalities $\mu_0 = s_{0k}\mu_k$, $k = 1, 2$, show that the divisor (μ_0) is equal to $B - 2\pi^* l_0$. Then, obviously, $K_S \sim -3\pi^* l_0 + 2B = \pi^* l_0 + (\mu_0^2) \sim \pi^* l_0$ and $K_S^2 = 3l_0^2 = 3$. Since $\pi^* l \in |K|$ for every line $l \subset \mathbf{P}^2$, we have $\dim_{\mathbf{C}} |K_S| \geq 2$, i.e. $p_g \geq 3$. If S is a minimal surface then a theorem of M. Noether (see [6], §X) guarantees the inequality $3 \geq 2p_g - 3$, i.e. $p_g \leq 3$; therefore $p_g = 3$. Thus, assume that S is not minimal. Then on S there exists an irreducible rational curve E with $-1 = (E, K_S) = (E, \pi^* l_0)$. But the generic element of the linear system which gives the map π is irreducible by the first Bertini theorem, and it can always be chosen not equal to the curve E, which is a contradiction.

The next theorem establishes a relation between Theorems 2.1 and 1.1 on the one hand, and Propositions 3.1 and 3.2 on the other.

THEOREM 3.1. *A surface S with invariants $p_g = 3$ and $K^2 = 3$ of type* I *with an ample canonical class is a CGC-surface if and only if in the presentation of Theorem* 2.1 *we have $f_2 = f_4 = 0$ (see* (2)).

PROOF. The "if" part is trivial. Let S be a CGC-surface. Then we can assume that $S \subset O_{\mathbf{P}^2}(2)$, as shown in Proposition 3.1. The system $\{\mu_i\}$ of fiber coordinates (see the proof of Proposition 3.2) defines a global section μ of the sheaf $O(2K_S)$, and $(\mu) = B$. Moreover in $H^0(S, O(6K_S))$ we obviously have $\mu^3 = G(x_0 : x_1 : x_2)$. It is now clear that if in the proof of Theorem 2.1 we set $C = B$, then on T we obtain $\xi^2 = \mu$ and $\xi^6 = G(x_0 : x_1 : x_2)$, i.e. the equation of the surface X ($X \cong S$) is $f_6(x_0 : x_1 : x_2) - x_3^6 = 0$. where f_6 is an irreducible and smooth form of degree 6.

PROPOSITION 3.3. *Two CGC-surfaces S' and S'' are isomorphic if and only if their branch loci Λ' and Λ'' are projectively equivalent in $\mathbf{P}^2$.*

PROOF. The "if" part is trivial because of Theorem 3.1. Let S' and S'' be isomorphic CGC-surfaces and let $\varphi\colon S' \to S''$ be one of the isomorphisms. φ induces an isomorphism of $\mathbf{C}$-linear spaces

$$\varphi^*\colon H^0(S'', O(K_{S''})) \to H^0(S', O(K_{S'})).$$

The canonical maps $\pi'\colon S' \to \mathbf{P}^2$ and $\pi''\colon S'' \to \mathbf{P}^2$ are given by a choice of bases:

$$H^0(S', O(K_{S'})) = \mathbf{C}\sigma_0' \oplus \mathbf{C}\sigma_1' \oplus \mathbf{C}\sigma_2', \qquad H^0(S'', O(K_{S''})) = \mathbf{C}\sigma_0'' \oplus \mathbf{C}\sigma_1'' \oplus \mathbf{C}\sigma_2'',$$

and the sections $\varphi^*(\sigma_i)$ also form a basis of the space $H^0(S', O(K_{S'}))$. Then there exists a nonsingular linear transformation α such that $\varphi^*(\sigma_i'') = \alpha(\sigma_i')$, i.e. the diagram

$$\begin{array}{ccc} S' & \xrightarrow{\varphi} & S'' \\ \pi' \downarrow & & \downarrow \pi'' \\ \mathbf{P}^2 & \xrightarrow{\alpha} & \mathbf{P}^2 \end{array}$$

is commutative. Since the branch locus Λ' of the cover π' is characterized set-theoretically as $y' \in \Lambda' \Leftrightarrow \#\pi^{-1}(y) = 1$, it is obvious that $\alpha\Lambda' = \Lambda''$.

§4. Moduli

The following theorem was first proved by Horikawa in [8], §5. The proof given here uses the presentation of the surface S from Theorem 2.1, and we think it is simpler and more natural.

THEOREM 4.1. *Let S be a minimal surface with $p_g = 3$ and $K^2 = 3$ of type* I *and suppose that its canonical class K is ample. Then the number of moduli exists and is equal to* 34. *The moduli space is nonsingular at the point S.*

PROOF. We start by proving

LEMMA 4.1. *Under the conditions of Theorem* 4.1, $H^0(S, \Theta) = 0$ *and* $H^2(S, \Theta) = 0$, *where Θ is a locally free sheaf corresponding to the tangent bundle.*

PROOF. By Serre duality we have

$$\dim_{\mathbf{C}} H^0(S, \Theta) = \dim_{\mathbf{C}} H^2(S, \Omega_S^1(K)) = \dim_{\mathbf{C}} H^0(S, \Omega_S^1(-K)).$$

Let K be an irreducible nonsingular curve from the canonical system $|K|$. The exact sequence

$$0 \to \Omega_S^1(-K) \to \Omega_S^1 \to \Omega_S^1(K) \to 0$$

induces the cohomology sequence

$$0 \to H^0(S < \Omega_S^1(-K)) \to H^0(S < \Omega_S^1) \to \cdots.$$

But $\dim_{\mathbf{C}} H^0(S, \Omega_S^1) = q =$ the irregularity of the surface $S = 0$ (see [6], §X, Theorem 13). Therefore $H^0(S, \Omega_S^1(-K)) = 0$. Let us calculate $\dim_{\mathbf{C}} H^2(S, \Theta)$. Using Serre duality again, we have

$$\dim_{\mathbf{C}} H^2(S, \Theta) = \dim_{\mathbf{C}} H^0(S, \Omega_S^1(K)).$$

Now let us use Theorem 1.1 and the presentation of the surface from Theorem 2.1. In the notation of §1 we have

$$H^0(X, \Omega_X^1(K)) = H^0(X, \Omega_X^1([H_0^*])) \cong \eta^* H^0(X, \Omega_X^1([H_0^*])) \subset H^0(Y, \Omega_Y^1([H_0]))$$

(here $X \cong S$). Let us prove that $H^0(Y, \Omega_Y'([H_0])) = 0$. We have the Severi exact sequence (see [3], §15):

$$0 \to \Omega_{\mathbf{P}^3}'^1([H_0]) \to \Omega_{\mathbf{P}^3}^1([H_0]) \to \Omega_Y^1([H_0]) \to 0$$

and the corresponding cohomology sequence

$$\begin{aligned} 0 &\to H^0(\mathbf{P}^3, \Omega_{\mathbf{P}^3}'^1([H_0])) \to H^0(\mathbf{P}^3, \Omega_{\mathbf{P}^3}^1([H_0])) \\ &\to H^0(Y, \Omega_Y^1([H_0])) \to H^1(\mathbf{P}^3, \Omega_{\mathbf{P}^3}'^1([H_0])) \to \cdots. \end{aligned} \tag{5}$$

Moreover the Poincaré exact sequence ([3], §15)

$$0 \to \Omega_{\mathbf{P}^3}^1(-5[H_0]) \to \Omega_{\mathbf{P}^3}'^1([H_0]) \to O_Y(-5[H_0]) \to 0$$

shows that the sequence

$$0 \to H^1(\mathbf{P}^3, \Omega^1_{\mathbf{P}^3}(-5[H_0])) \to H^1(\mathbf{P}^3, \Omega'^1_{\mathbf{P}^3}([H_0])) \to H^1(Y, O_Y(-5[H_0])) \to \cdots \tag{6}$$

is exact. By a theorem of Ramanujan ([6], §III, the corollary to Theorem A) we have $H^1(Y, O_Y(-5[H_0])) = 0$. On the other hand, by Bott's result (see [12], §5) the groups $H^0(\mathbf{P}^3, \Omega^1_{\mathbf{P}^2}([H_0]))$ and $H^1(\mathbf{P}^3, \Omega^1_{\mathbf{P}^3}(-5[H_0]))$ are zero; then from (5) and (6) we obtain $H^0(Y, \Omega^1_Y([H_0])) = 0$. Lemma 4.1 is proved.

Lemma 4.1 and the Kodaira-Spencer-Nirenberg theorem ([9], §1) imply that the number of moduli $m(S)$ exists and is equal to $\dim_{\mathbf{C}} H^1(S, \Theta)$. By the Riemann-Roch-Hirzebruch theorem we have (see [10], §11)

$$\dim_{\mathbf{C}} H^1(S, \Theta) = 10(p_g - q + 1) - 2K^2 + \dim_{\mathbf{C}} H^0(S, \Theta) + \dim_{\mathbf{C}} H^2(S, \Theta) = 34.$$

Moreover, the moduli space is nonsingular at the point S. Theorem 4.1 is proved.

The results of §3 show that the moduli space $M^{\mathrm{CGC}}_{3,3}$ of all CGC-surfaces is the quotient $U/PGL(2)$, where U is the set of smooth plane curves of degree 6 (U is an open subset of $\mathbf{P}^{27}$). Mumford (see [1], Chapter 4, §2) proved that the elements of U are stable points with respect to the action of $PGL(2)$ and therefore the quotient has an algebraic structure.

A. N. Todorov ([2], Chapter II), proved that this quotient has a homogeneous representation

$$U/PGL(2) \cong \Gamma \backslash SO(2,19)/SO(2) \times SO(19),$$

where $\Gamma = SO(2, 19; \mathbf{Z})$. It is clear now that the moduli space $M^{\mathrm{CGC}}_{3,3}$ is an irreducible algebraic variety of dimension 19.

BIBLIOGRAPHY

1. David Mumford, *Geometric invariant theory*, Springer-Verlag, Berlin, and Academic Press, New York, 1965.
2. A. N. Todorov, *Jordan cells of monodromies and applications to the moduli of K3 surfaces*, Dissertation, Moscow State Univ., Moscow, 1975. (Russian)
3. S. S. Chern, *Complex manifolds*, Inst. Fis. e Mat. Univ. Recife, Recife, 1959.
4. I. R. Shafarevich et al., *Algebraic surfaces*, Trudy Mat. Inst. Steklov. **75** (1965); English transl., Proc. Steklov Inst. Math. **75** (1965).
5. I. R. Shafarevich, *Basic algebraic geometry*, "Nauka", Moscow, 1972; English transl., Springer-Verlag, 1974.
6. E. Bombieri, *Canonical models of surfaces of general type*, Inst. Hautes Études Sci. Publ. Math. No. 42 (1973), 171–219.
7. Robin Hartshorne, *Algebraic geometry*, Springer-Verlag, 1977.
8. Eiji Horikawa, *Algebraic surfaces of general type with small* c_1^2. II, Invent. Math. **37** (1976), 121–155.
9. K. Kodaira, L. Nirenberg, and D. C. Spencer, *On the existence of deformations of complex analytic structures*, Ann. of Math. (2) **68** (1958), 450–459.
10. K. Kodaira and D. C. Spencer, *On deformations of complex analytic structures*. I, Ann. of Math. (2) **67** (1958), 328–401.
11. Joseph Steenbrink, *Intersection form for quasi-homogeneous singularities*, Compositio Math. **34** (1977), 211–223.
12. John J. Wavrik, *Deformations of Banach coverings of complex manifolds*, Amer. J. Math. **90** (1968), 926–960.

Translated by A. MARTSINKOVSKY

Amer. Math. Soc. Transl.
(2) Vol. **132**, 1986

Surfaces with $p_g = 3$ and $K^2 = 3$. Part II

VALENTIN V. ILIEV

Abstract. The author gives a description of the surfaces of type I with $p_g = 3$ and $K^2 = 3$ (the canonical class is not required to be ample). They are the minimal resolutions of the rational double points of their normal canonical models which are realized as surfaces of degree 6 in the weighted 3-dimensional projective space of type $(1,1,1,2)$. The two-canonical map is biholomorphic mod $\mathcal{E}$. Two surfaces of this type are isomorphic if and only if their normal canonical models are G-equivalent, where G is a 15-dimensional algebraic group.

This paper is a direct continuation of [2] and also uses the same notation.

§5. A description of the surfaces of type I with $p_g = 3$ and $K^2 = 3$

THEOREM 5.1. *Let S be a minimal surface of type* I *with $p_g = 3$ and $K^2 = 3$. Then its normal canonical model embeds isomorphically in $\mathbf{P}^3(1,1,1,2)$ as the set X consisting of all zeros of a quasihomogeneous polynomial of type $(1,1,1,2)$ of degree* 6 *satisfying condition* C) *from* §1. *Moreover, the quotient map $\Phi^{\sigma}_{\pi^*K}\colon S \to X$ is the minimal resolution of singularities of X.*

PROOF. Lemmas 2.1–2.11 all remain true even if the canonical class is not ample. In particular, Lemma 2.11 implies that in the commutative diagram

$$\begin{array}{ccc} T & \xrightarrow{\Phi_{\pi^*K}} & Y \\ \pi\downarrow & & \downarrow\eta \\ S & \xrightarrow{\Phi^{\sigma}_{\pi^*K}} & X \end{array}$$

the map Φ_{π^*K} is biregular mod $(\bigcup_1^t E_i)$ and is the minimal resolution of the rational double points of Y. If $Y = v(F)$, then

$$X = v(F^*) \subset \mathbf{P}^3(1,1,1,2),$$

1980 *Mathematics Subject Classification* (1985 *Revision*). Primary 14J10, 14J25.
Translation of Serdica **7** (1981), 390–395; MR **83i**: 14026 b.

where $F(x_0{:}x_1{:}x_2{:}x_3) = F^*(x_0, x_1, x_2, x_3^2)$. Thus we have an isomorphism

$$X = \operatorname{Proj}(\mathbf{C}[w_0, w_1, w_2, w_3]/(F^*)).$$

On the other hand, Lemmas 2.3 and 2.10 imply that

$$\bigoplus_{m\geq 0} H^0(S, O(mK)) = \mathbf{C}[\lambda_0, \lambda_1, \lambda_2, \mu]$$

and $F^*(\lambda_0, \lambda_1, \lambda_2, \mu) = 0$ is the only algebraic relation between $\lambda_0, \lambda_1, \lambda_2$, and μ, i.e.

$$\mathbf{C}[\lambda_0, \lambda_1, \lambda_2, \mu] \cong \mathbf{C}[w_0, w_1, w_2, w_3]/(F^*)$$

and X is the normal canonical model of the surface S.

Since $\Gamma = \Phi^{-1}_{\pi^*K}(H_3)$ and $2\Gamma \sim K_T$, we have $(\Gamma, E_i) = 0$, i.e. the ramification curve H_3 of the covering η does not contain points from $\operatorname{Sing}(Y)$ (we have $\bigcup_1^t E_i = \Phi^{-1}_{\pi^*K}(\operatorname{Sing}(Y))$). Hence $\operatorname{Sing}(Y) = \eta^{-1}(\operatorname{Sing}(X))$. The involution σ induces a permutation σ on the set $\{E_1, \ldots, E_t\}$ which is again an involution. Let $\sigma = \tau_1 \cdots \tau_s$ be its decomposition into disjoint cycles. Obviously the length of τ_k is at most 2. On the other hand, if $\Phi_{\pi^*K}(E_i) = y$ then $\Phi_{\pi^*K}(\sigma E_i) = \sigma(y) \neq y$. In particular $E_i \cap \sigma E_i = \varnothing$ and therefore the length of τ_k is at least 2, i.e. the length of τ_k is 2 for each $k = 1, \ldots, s$. This means that every cycle τ_k is transposition, $t = 2s$, and if $\pi(\{E_1, \ldots, E_t\}) = \{E'_i, \ldots, E'_s\}$ then $\pi^* E'_k = E_i + \sigma E_i$, where $\tau_k(E_i) = \sigma E_i$ and $\tau_k(\sigma E_i) = E_i$. Moreover $\Phi^{\sigma}_{\pi^*K}(E'_k) = x$, where $x = \eta(y)$, and the restriction

$$\Phi_{\pi^*K}{:}\, S - \bigcup_{k=1}^{s} E'_k \to X - \operatorname{Sing}(X)$$

is an isomorphism. Hence $\Phi^{\sigma}_{\pi^*K}{:}\, S \to X$ is the minimal resolution of the rational double points of X. The equalities $E'^2_k = -2$ and the rationality of the curves E'_k follow from this, although one can derive them directly:

$$E'^2_k = \tfrac{1}{2}(\pi^* E'_k)^2 = \tfrac{1}{2}(E_i + \sigma E_i)^2 = -2, \qquad (K, E'_k) = 0,$$

which implies that the virtual genus of E'_k is 0, i.e. E'_k is a rational curve. Theorem 5.1 is proved.

THEOREM 5.2. *The 2-canonical map* $\Phi_{2K}{:}\, S \to \mathbf{P}^6$ *is biholomorphic* mod $(\bigcup_1^s E'_k)$.

PROOF. We have $\pi^{-1}(S - \bigcup_1^s E'_k) = T - \bigcup_1^t E_i$, and the statement follows from Lemmas 2.7, 2.8, and 2.11, and the obvious fact that Φ_{2K} contracts connected components of $\bigcup_1^s E'_k$ to points.

REMARK 1. Since the restriction $\Phi_{\pi^*K}{:}\, \Gamma \to H_3$ is an isomorphism, H_3 is an irreducible smooth curve.

THEOREM 5.3. *Let* $X = v(F^*) \subset \mathbf{P}^3(1,1,1,2)$, *where* $F^*(w_0, w_1, w_2, w_3)$ *is an irreducible quasihomogeneous polynomial of type* $(1,1,1,2)$ *of degree* 6 *with the extra conditions that* $F = F^*(x_0, x_1, x_2, x_3^2)$ *is of the form* (2) *and satisfies*

C) *from* §1 *and* Sing(x) *consists of rational double points. Then the minimal resolution of these singularities is a surface of type* I *with* $p_g = 3$ *and* $K^2 = 3$.

PROOF. Let $Y = v(F) \subset \mathbf{P}^3$, and let $\eta: Y \to X$ be the canonical 2-fold cover ramified along H_3, for which by C) we have $H_3 \cap \mathrm{Sing}(Y) = \varnothing$. Then obviously $\eta^{-1}(\mathrm{Sing}(X)) = \mathrm{Sing}(Y)$ and Y has only rational double points. Let $\psi': T \to Y$ be the minimal resolution of these singularities and $\psi = j \circ \psi'$, where $j: Y \to \mathbf{P}^3$ is the natural embedding. If $\mathcal{L}$ denotes $\psi^* O_{\mathbf{P}^3}(1)$ and D denotes $\psi^*(H_0)$, then $\mathcal{L} = O([D])$. After the change of coordinates

$$\begin{pmatrix} x_0' \\ x_1' \\ x_2' \end{pmatrix} = A^{-1} \begin{pmatrix} x_0 \\ x_1 \\ x_2 \end{pmatrix},$$

$$x_3' = x_3, \qquad A \in GL_3(\mathbf{C}),$$

(which induces a change of "coordinates" in $\mathbf{P}^3(1,1,1,2)$) we may assume that $H_0 \cap \mathrm{Sing}(Y) = \varnothing$ and then, obviously, $D^2 = H_0^2 = 6$. Furthermore, despite the presence of rational double points on Y, we can apply a construction from [4], chapter III, §5.4, which shows that K_Y is a locally principal divisor on Y and $K_Y \sim 2H_0$. With the aid of Theorem 2.7 we obtain from [5] that $K_T \sim \psi^*(2H_0) = 2D$. It follows immediately that T is a minimal surface. Indeed, if Λ is an exceptional curve of the first kind on T, then $-1 = (K_T, \Lambda) = 2(D, \Lambda)$, a contradiction.

By means of the birational isomorphism ψ' the involution σ lifts to a birational involution σ on T, which turns out to be biregular since T is a minimal surface. Moreover the smooth irreducible curve $\Gamma = \psi^*(H_3)$ is fixed by σ. The automorphisms σ and $\sigma^2 = \mathrm{id}$ form the group $\mathbf{Z}_2$. Let $S = T/\mathbf{Z}_2$, and let $\pi: T \to S$ be the natural projection with the ramification curve Γ and the branch locus $C = \pi(\Gamma)$. The 2-form $\omega = y_3\omega_3$ is σ-invariant on Y, and $(\omega)|_\gamma = H_0 + H_3$. Obviously ω is a σ-invariant 2-form on T, and $(\omega)|_T = D + \Gamma$, i.e. (Lemma 1.1) ω is a regular 2-form on S and $(\omega)|_S = K$ with $\pi^*K = D$. If

$$\lambda_0 = \psi^*(x_0), \quad \lambda_1 = \psi^*(x_1), \quad \lambda_2 = \psi^*(x_2), \quad \xi = \psi^*(x_3)$$

then, since $H^0(Y, O([H_0])) = \mathbf{C}x_0 \oplus \mathbf{C}x_1 \oplus \mathbf{C}x_2 \oplus \mathbf{C}x_3$ we have

$$H^0(T, O(\pi^*K)) = \mathbf{C}\lambda_0 \oplus \mathbf{C}\lambda_1 \oplus \mathbf{C}\lambda_2 \oplus \mathbf{C}\xi$$

and $\lambda_i^\sigma = \lambda_i$, $\xi^\sigma = -\xi$, i.e. $H^0(S, O(K)) = \mathbf{C}\lambda_0 \oplus \mathbf{C}\lambda_1 \oplus \mathbf{C}\lambda_2$ and, in particular, $p_g(S) = 3$. Moreover we have $K_T \sim \pi^*K + \Gamma$ and $K_T \sim 2D \sim 2\Gamma$; thus $\pi^*K \sim \Gamma$ and $K^2 = \frac{1}{2}(\pi^*K)^2 = \frac{1}{2}\Gamma^2 = \frac{1}{2}D^2 = 3$. Repeating with minor modifications the arguments of Step 4 from §1, we see that S is a minimal surface.

Finally it is easy to see that the 2-form $\omega = y_3^4\omega_3^{\otimes 2}$ of weight 2 is carried over to S and $(\omega)|_S = C$, i.e. $C \in |2K|$.

Thus we are in the situation of the proof of Theorem 5.1. This implies that, in the notation $\Phi_{\pi^*K} = \psi$, $\Phi_{\pi^*K}^\sigma: S \to X$ is the minimal resolution of X. Theorem 5.3 is proved.

Theorems 5.1 and 5.3 correspond to Theorems 2.1 and 1.1 from Part I of this paper and complete the description of the surfaces of type I.

§6. Towards a global moduli space

Let the polynomial ring $\mathbf{C}[w_0, w_1, w_2, w_3]$ be graded by conditions $\deg w_0 = \deg w_1 = \deg w_2 = 1$ and $\deg w_3 = 2$, and let $\mathbf{C}[w, w_3] = \bigoplus_{m\geq 0} M_m$ be the corresponding grading. Here M_m is a $\mathbf{C}$-linear space of quasihomogeneous polynomials of type $(1,1,1,2)$ of degree m. Also set $w = (w_0, w_1, w_2)$ for short. We have $M_0 = \mathbf{C}$, $M_1 = (w_0, w_1, w_2)$, $M_2 = (w_i w_j, 0 \leq i \leq j \leq 2, w_3)$, etc. If φ is a graded automorphism of degree 0 of the ring $\mathbf{C}[w,w_3]$, then its restriction φ to M_m is a $\mathbf{C}$-linear automorphism of the space M_m. Let

$$\begin{pmatrix} \varphi(w_0) \\ \varphi(w_1) \\ \varphi(w_2) \end{pmatrix} = A^{-1} \begin{pmatrix} w_0 \\ w_1 \\ w_2 \end{pmatrix},$$

$$\varphi(w_3) = c w_3 - H(A^{-1} w),$$

where $A \in GL_3(\mathbf{C})$, $H = \sum_{0\leq i \leq j \leq 2} h_{ij} w_i w_j$, $h_{ij} \in \mathbf{C}$, and $c \in \mathbf{C}^*$. We may assume that $H \in S_3(\mathbf{C})$, where $S_3(\mathbf{C})$ is the set of symmetric matrices of order 3. A triplet (A, H, c) from $GL_3(\mathbf{C}) \times S_3(\mathbf{C}) \times \mathbf{C}^*$ completely determines the corresponding φ. Thus we have a bijection

$$\operatorname{Grad}\operatorname{Aut}_0 \mathbf{C}[w, w_3] \to GL_3(\mathbf{C}) \times S_3(\mathbf{C}) \times \mathbf{C}^*$$

$$\varphi \mapsto (A, H, c)$$

and it endows the algebraic variety $G' = GL_3(\mathbf{C}) \times S_3(\mathbf{C}) \times \mathbf{C}^*$ with the structure of an algebraic group with the group law

$$(A, H, c)(B, K, d) = (AB, d\,{}^t\!BHB + K, cd),$$

where ${}^t\!B$ denotes the transpose of B. It is easy to see that the center of G' is $Z(G') = \{(D(t^{-1})), 0, t^2 | t \in \mathbf{C}^*\} \cong \mathbf{C}^*$ ($D(t)$ is the scalar matrix with element t on the diagonal) and that it consists of the elements of G' which correspond to the graded automorphisms of degree 0 inducing the identity map on the scheme $\mathbf{P}^3(1,1,1,2) \cong \operatorname{Proj} \mathbf{C}[w, w_3]$. Let G denote the algebraic group $GL_3(\mathbf{C}) \times S_3(\mathbf{C})$ with the group law

$$(A, H)(B, K) = (AB, {}^t\!BHB + K).$$

We have the natural embedding $G \to G'$, $(A, H) \mapsto (A, H, 1)$, and the composition $G \to G' \to G'/Z(G')$ is surjective with kernel $\{(\pm I, 0)\} \cong \mathbf{Z}_2$. By means of this embedding we can define a natural action of G on the graded ring $\mathbf{C}[w, w_3] = \bigoplus_{m \geq 0} M_m$: if $\varphi_{(A,H)}$ denotes the automorphism from $\operatorname{Grad}\operatorname{Aut}_0 \mathbf{C}[w, w_3]$ corresponding to the element $(A, H, 1) \in G'$ and $F^* \in \mathbf{C}[w, w_3]$, then $(A, H)F^* = \varphi_{(A,H)}(F^*)$. On the scheme $\mathbf{P}^3(1,1,1,2)$ the group acts in the dual way.

If S is a surface of type I, then, because of Theorem 5.1, the canonical homomorphism $\mathbf{C}[w, w_3] \to \mathbf{C}[\lambda, \mu] \to 0$, where $\lambda = (\lambda_0, \lambda_1, \lambda_2)$, induces an embedding $X \subset \mathbf{P}^3(1,1,1,2)$, and obviously two surfaces S' and S'' are isomorphic if and only if their normal canonical models $X' = \imath(F_1^*)$ and $X'' = v(F_2^*)$ are isomorphic. The following theorem justifies the existence of this section.

THEOREM 6.1. *Two surfaces S' and S'' of type* I *with $p_g = 3$ and $K^2 = 3$ are isomorphic if and only if their normal canonical models $X' = v(F_1^*)$ and $X'' = v(F_2'')$ in $\mathbf{P}^3(1,1,1,2)$ are G-equivalent.*

PROOF. The "if" part is trivial. Let S' and S'' be isomorphic. An isomorphism between them induces a graded isomorphism of degree 0 of their polycanonical rings:

$$\varphi\colon \mathbf{C}[\lambda', \mu'] \underset{\substack{\lambda' \mapsto \lambda'' \\ \mu' \mapsto \mu''}}{\leftrightarrow} \mathbf{C}[\lambda'', \mu''],$$

which is obviously determined by the data $\lambda'' = A^{-1}\lambda'$ and $\mu'' = c\mu' - H(A^{-1}\lambda')$, where $A \in GL_3(\mathbf{C})$, $H \in S_3(\mathbf{C})$, and $c \in \mathbf{C}^*$. The same data define a graded automorphism φ of degree 0 of $\mathbf{C}[w', w_3']$ such that the diagram

$$\begin{array}{ccc} \mathbf{C}[w', w_3'] & \overset{\varphi}{\to} & \mathbf{C}[w'', w_3''] \\ \downarrow & & \downarrow \\ \mathbf{C}[\lambda', \mu'] & \overset{\varphi}{\to} & \mathbf{C}[\lambda'', \mu''] \\ \downarrow & & \downarrow \\ 0 & & 0 \end{array} \tag{7}$$

is commutative. The dual diagram is also commutative:

$$\begin{array}{ccc} \mathbf{P}^3(1,1,1,2) & \overset{\varphi^*}{\leftarrow} & \mathbf{P}^3(1,1,1,2), \\ \uparrow & & \uparrow \\ X' & \overset{\varphi^*}{\leftarrow} & X'' \end{array} \tag{8}$$

Note that the vertical arrows are closed immersions. The prime ideals (F_1^*) and (F_2^*) are the kernels of the surjective homomorphisms (7), and obviously $\varphi^*((F_2^*)) = (F_1^*)$, i.e. $\varphi^{-1}((F_2^*)) = (F_1^*)$ or $(F_2^*) = \varphi((F_1^*)) = (\varphi F_1^*)$. If $c \neq 1$ we can take the composition of φ with the automorphism corresponding to the element $(D(c^{1/2}), 0, c^{-1})$ from the center of G'. This does not change the closed immersions (8), and moreover φF_1^* again has highest coefficient 1, i.e. $\varphi F_1^* = F_2^*$ or

$$F_2^*(w', w_3') = F_1^*(w'', w_3'') = F_1^*(A^{-1}w', w_3' - H(A^{-1\prime}w)).$$

It is now clear that $X' = \varphi^*_{(A,H)}(X'')$, i.e. X' and X'' are G-equivalent. Theorem 6.1 is proved.

Let V_1 be the set of quasihomogeneous polynomials of type $(1,1,1,2)$ and degree 6 which have only rational double points and satisfy the following condition: if

$$F^*(w, w_3) = f_6(w) + f_4(w)w_3 + f_2(w)w_3^2 + cw_3^3, \tag{9}$$

then $c \neq 0$ and $\{f_6(w) = 0\}$ is a smooth plane curve (such polynomials exist; e.g., the polynomial

$$F^*(w, w_3) = (1/54)w_0^6 - w_1^6 - w_2^6 - (\tfrac{1}{2}w_0^2 - w_1^2 + w_2^2)w_3^2 - w_3^3$$

has only one rational double point with "coordinates" $(1,0,0,\pm\sqrt{-1}/\sqrt{3}))$. We can assume that $c=-1$. Then V_1 is an open set in $\mathbf{C}^{49}=\mathbf{P}^{49}-\{c=0\}$ (here $\mathbf{P}^{49}=\mathbf{P}(M_6)$). G acts on $\mathbf{P}^{49}=\mathbf{P}(M_6)$ as follows:

$$(A,H)F^*(w,w_3)=F^*(A^{-1}w,w_3-H(A^{-1}w)).$$

Let $V=\bigcup_{(A,H)\in G}(A,H)V_1$. Then V is G-invariant, and Theorem 6.1 shows that the set consisting of all surfaces of type I up to isomorphism is in one-to-one correspondence with the quotient V/G. The moduli problem is to endow this set with some algebraic structure so that the canonical map $V\to V/G$ is a morphism in the corresponding category.

LEMMA 6.1. *If for F^* from (9) the polynomial $3f_4(w)+f_2^2(w)$ is smooth or if $f_2(w)=f_4(w)=0$, then the stabilizer $S(F^*)=\{(A,H)\in G|(A,H)F^*=F^*\}$ is finite.*

PROOF. By the Taylor formula,

$$\begin{aligned}(A,H)F^* &= F^*(A^{-1}w,w_3-H(A^{-1}w))\\ &= F^*(A^{-1}w,-H(A^{-1}w))+(\partial F^*/\partial w_3)(A^{-1}w,-H(A^{-1}w))w_3\\ &\quad+\tfrac{1}{2}(\partial^2F^*/\partial w_3^2)(A^{-1}w,-H(A^{-1}w))w_3^2-w_3^3,\end{aligned}$$

and $(A,H)\in S(F^*)$ means that

$$\begin{aligned}f_2(w)&=\tfrac{1}{2}(\partial^2F^*/\partial w_3^2)(A^{-1}w,-H(A^{-1}w)),\\ f_4(w)&=(\partial F^*/\partial w_3)(A^{-1}w,-H(A^{-1}w)),\\ f_6(w)&=F^*(A^{-1}w,-H(A^{-1}w)).\end{aligned}$$

The first equation gives $H(A^{-1}w)=\frac{1}{3}(f_2(w)-f_2(A^{-1}w))$, and the second equation shows that the degree 4 polynomial $3f_4(w)+f_2^2(w)$ is A-invariant. If $f_2(w)=f_4(w)=0$, then the third equation shows that f_6 is A-invariant. In both cases we can use the classical result that if a smooth homogeneous polynomial of degree greater than 2 is A-invariant, then there are only a finite number of such A in $GL_3(\mathbf{C})$ (see [6]). Lemma 6.1 is proved.

Let V_1' be the open subset of V_1 consisting of F^* for which the polynomial $3f_4(w)+f_2^2(w)$ is smooth, and let $V'=\bigcup_{(A,H)\in G}(A,H)_1'$. The set V' is G-invariant, and if $O(F^*)$ is the orbit of F^* then $\dim O(F^*)=\dim G=15$ ($O(F^*)\cong G/S(F^*)$ as algebraic varieties). Moreover the lemma on closed orbits (see [1], Chapter IV, §3) shows that $O(F^*)$ is closed in V', i.e. we have

LEMMA 6.2. *The action of G on the quasiprojective variety V' is closed.*

LEMMA 6.3. *The group G is a semidirect product of $GL_3(\mathbf{C})$ and $S_3(\mathbf{C})$ with respect to the representation*

$$\begin{aligned}GL_3(\mathbf{C})&\to \operatorname{Aut}S_3(\mathbf{C}),\\ A&\mapsto (H\to {}^tAHA).\end{aligned}$$

PROOF. This follows from the definition of the group operation in G. Note that by means of the closed embedding $H\to(I,H)$ (respectively $A\to(A,0)$)

we can identify $S_3(\mathbf{C})$ $(GL_3(\mathbf{C}))$ with the normal subgroup (closed subgroup) of G, and the quotient group $G/S_3(\mathbf{C})$ is isomorphic to $GL_3(\mathbf{C})$. Let V'' denote the set of all F^* such that $f_2(w) = f_4(w)$. Those are the surfaces described in §3. Then $GL_3(\mathbf{C}) = \{(A, 0) \in G\}$ coincides with the set of elements $(A, H) \in G$ for which $F^* \in V''$ implies $(A, H)F^* \in V''$. It is now clear that there is an action of $GL_3(\mathbf{C})$ of V'', and Lemma 6.1 shows that every element $F^* \in V''$ has finite stabilizer. In this context Theorem 6.1 coincides with Proposition 3.3: the quotient $V''/GL_3(\mathbf{C})$ is isomorphic to the quotient $U/PGL(2)$ from the end of §4.

REFERENCES

1. Jean A. Dieudonné and James B. Carrell, *Invariant theory, old and new*, Advances in Math. **4** (1970), 1–80; reprinted as a separate book, Academic Press, 1971.

2. Valentin V. Iliev, *Surfaces with $p_g = 3$ and $K^2 = 3$. Part* I, Serdica **6** (1980), 352–360; English transl. in this volume.

3. S. S. Chern, *Complex manifolds*, Inst. Fis. e Mat. Univ. Recife, Recife, 1959.

4. I. R. Shafarevich, *Basic algebraic geometry*, "Nauka", Moscow, 1972; English transl., Springer-Verlag, 1974.

5. Michael Artin, *Some numerical criteria for contractibility of curves on algebraic surfaces*, Amer. J. Math. **84** (1962), 485–496.

6. Peter Orlik and Louis Solomon, *Singularities*. II : *Automorphisms of forms*, Math. Ann. **231** (1977/78), 229–240.

Translated by A. MARTSINKOVSKY

Amer. Math. Soc. Transl.
(2) Vol. **132**, 1986

A Formula for the Characters of Irreducible Finite-Dimensional Representations of Lie Superalgebras of Series C

D. A. LEĬTES

This note is a direct continuation of [1]. It uses the definitions and notation of [1], replacing Gl and sl by $\operatorname{osp}(2,2n) = c(n-1)$. In particular, the definition of a tensor field of type V on $(G_{-1})^*$, where $G = c(n-1)$ and V is a G_0-module, is given analogously to [1].

THEOREM 1. *Let V_χ be an irreducible G-module with highest weight χ, where the highest vector is even. Then*

$$\operatorname{ch} V_\chi = \left\{ \sum_{w\in W} \operatorname{sgn} w \cdot w \left[e^{\chi+P} \prod_{(\chi+P,\varphi_i)\neq 0} (1+\varepsilon e^{\varphi_i}) \right] \right\} \Big/ \left(\sum_{w\in W} \operatorname{sgn} w e^{wP} \right).$$

Let ε_i be the weight of the ith fundamental representation of the Lie algebra $\operatorname{sp}(2n)$. We put

$$\Sigma_i(a_1\varepsilon_1 + \cdots + a_n\varepsilon_n) = [\Omega^i(a_1\varepsilon_1 + \cdots + a_n\varepsilon_n)]^* = T(a_1, \ldots, a_n; a_i + 2n - i + 1),$$

where the highest weight χ is given by its values on the standard basis [2] of a Cartan subalgebra in $\operatorname{sp}(2n)$ and on an element z of the center of G_0 for which $[z, G_1] = iG_i$.

THEOREM 2. *Let V_1 and V_2 be irreducible G_0-modules and let $c\colon T(V_1) \to T(V_2)$ be a G-invariant differential operator. Then $T(V_1)$ and $T(V_2)$ are adjacent terms in the symmetric sequences (1)–(2^{n-1}), and c is proportional to the corresponding differential operator d_j of order $j = 1, 2, \ldots, n, 2n$.*

1980 *Mathematics Subject Classification* (1985 *Revision*). Primary 17A70; Secondary 17B70, 81C40.

Translation of C. R. Acad. Bulgare Sci. **33** (1980), 1053–1055; MR **82j**: 17020b.

$$
\begin{array}{c}
\cdots \xrightarrow{d_1} \Sigma_1(a_2+1, a_2, \ldots, a_n) \xrightarrow{d_1} \Sigma_1(a_2, a_2, \ldots, a_n) \\
\overset{d_2}{\swarrow} \\
\Sigma_2(a_2-1, a_2-1, a_3, \ldots, a_n) \xrightarrow{d_1} \cdots \xrightarrow{d_1} \Sigma_2(a_2-1, a_3, a_3, \ldots, a_n) \\
\overset{d_2}{\swarrow} \\
\cdots\cdots\cdots\cdots\cdots\cdots\cdots\cdots\cdots\cdots\cdots\cdots \\
\overset{d_2}{\swarrow} \\
\Sigma_n(a_2-1, \ldots, a_n-1, a_n-1) \xrightarrow{d_1} \cdots \xrightarrow{d_1} \Sigma_n(a_2-1, \ldots, a_n-1, 0) \\
\overset{d_2}{\swarrow} \\
\Omega^n(a_2-1, \ldots, a_n-1, 0) \xrightarrow{d_1} \cdots \xrightarrow{d_1} \Omega^n(a_2-1, \ldots, a_n-1, a_n-1) \\
\overset{d_2}{\swarrow} \\
\cdots\cdots\cdots\cdots\cdots\cdots\cdots\cdots\cdots\cdots\cdots\cdots \\
\overset{d_2}{\swarrow} \\
\Omega^1(a_2, a_2, \ldots, a_n) \xrightarrow{d_1} \Omega^1(a_2+1, a_2, \ldots, a_n) \xrightarrow{d_1} \cdots
\end{array}
\tag{1}
$$

Here (as in [1] and [3]) the top and bottom lines are infinite, and the $n-1$ intermediate lines are finite, while all the horizontals are assembled into a single concertina by the differential operator d_2. When any of the short lines containing Σ degenerate, then the lines containing the dual superspaces Ω also degenerate, while the differential operators going through the place where they would have been, and connecting all the lines into the concertina, have degree $2+r$, where r is the number of omitted lines.

$$
\begin{array}{c}
\cdots \xrightarrow{d_1} \Sigma_1(2\varepsilon_1) \xrightarrow{d_1} \Sigma_1(\varepsilon_1) \xrightarrow{d_1} \Sigma_1(0) \\
\overset{d_{2n}}{\swarrow} \\
\Omega_1(0) \xrightarrow{d_1} \Omega_1(\varepsilon_1) \xrightarrow{d_1} \cdots
\end{array}
\tag{2^{n-1}}
$$

REMARK. The sequence (2^{n-1}) is an ordinary sequence of integro-differential forms (see [1]).

THEOREM 3. *The sequences* (1)–(2^{n-1}) *are exact.*

THEOREM 4. *Every* $c(n-1)$*-module* $T(\chi)$ *has an irreducible* $c(n-1)$*-submodule* $\operatorname{irr} T(\chi)$. *If* $T(\chi)$ *is not contained in the sequences* (1)–(2^{n-1}), *then* $\operatorname{irr} T(\chi) = T(\chi)$. *Otherwise* $\operatorname{irr} T(\chi)$ *is the kernel of the corresponding operator.*

The modules $\operatorname{irr} T(\chi)$ and $\pi(\operatorname{irr} T(\chi))$ are pairwise nonequivalent and together exhaust all the irreducible finite-dimensional $c(n-1)$-modules.

By the same considerations as in [1], for the proof of Theorem 2 it suffices to describe the singular vectors of degree -1. We choose the following basis in $G = c(n-1)$: let

$$G_{-1} = \langle l_1, \ldots, l_n, l'_1, \ldots, l'_n \rangle, \qquad G_0 = \langle z, A_{ij}, B_{ij}, c_{ij} \rangle,$$

where $B_{ji} = B_{ij}$ and $c_{ji} = c_{ij}$, $1 \le i, j \le n$, and

$$G_1 = \langle s_1, \dots, s_n, s'_1, \dots, s'_n \rangle.$$

A matrix representation in $(2, 2n)$-dimensional superspace is given by the formula

$$\left(\begin{array}{cc|c|c} z & 0 & s_1 \dots s_n & s'_1 \dots s'_n \\ 0 & z & l_1 \dots l_n & l'_1 \dots l'_n \\ \hline l'_1 & S'_1 & & \\ \dots & \dots & A & B \\ l'_n & S'_n & & \\ \hline -l_1 & -S_1 & & \\ \dots & \dots & C & -A^t \\ l_n & S_n & & \end{array}\right)$$

Let $f_1 = l_1 v_1 + \dots + l_n v_n + l'_n v'_n + \dots + l'_1 v'_1$ be the highest singular vector (for the definition of a singular vector, see [3]). Then:

$s_1(f_1) = 0$ is the condition for singularity; and

$B_{ij}(f_1) = 0$ and $A_{kl}(f_1) = 0$, $k < l$, is the condition for being higher.

The consequences of these conditions (for convenience we have also computed $s_2(f_1), \dots, s'_n(f_1)$) are

$$\begin{gathered} 2c_{11}v_1 + \dots + (c_{1n} + c_{n1})v_n - (A_{11} + z)v'_1 - A_{21}v'_2 - \dots - A_{n1}v'_n = 0, \\ \dots\dots\dots\dots\dots\dots\dots\dots\dots\dots\dots\dots\dots\dots\dots \\ (c_{1n} + c_{n1})v_1 + \dots + 2c_{nn}v_n - A_{1n}v'_1 - \dots - (A_{nn} + z)v'_n = 0, \\ -(A_{11} - z)v_1 - \dots - A_{1n}v_n - 2B_{11}v'_1 - \dots - (B_{1n} + B_{n1})v_n = 0, \\ \dots\dots\dots\dots\dots\dots\dots\dots\dots\dots\dots\dots\dots\dots\dots \\ -A_{n1}v_1 - \dots - (A_{nn} - z)v_n - (B_{n1} + B_{1n})v'_1 - \dots - 2B_{nn}v'_n = 0, \\ B_{ij}v'_1 = 0, \dots, B_{ij}v'_i - v_j = 0, \dots, B_{ij}v'_j - v_i = 0, \dots, \qquad i < j \\ A_{ij}v_1 = 0, \dots, A_{ij}v_j - v_i = 0, \dots, A_{ij}v'_i - v'_j = 0, \dots, \qquad i < j. \end{gathered}$$

From these consequences we deduce the following facts:

If $v_1 = \dots = v_{k-1} = 0$, $v_k \neq 0$, *then*

$$\chi(v_k) = (a_k + 2n - k + 1; a_1, \dots, a_n),$$

$$\chi(f_1) = (a_k + 2n - k; a_1, \dots, a_k - 1, \dots, a_n),$$

and if $v_1 = \dots = v_n = v'_n = \dots = v'_{n-k+1} = 0$, *then*

$$\chi(v'_{n-k}) = (-a_k + k - 1; a_1, \dots, a_n),$$

$$\chi(f_1) = (-a_k + k - 2; a_1, \dots, a_k + 1, \dots, a_n),$$

where by $\chi(v)$ *we denote the weight of the vector* v.

The rest of the proof is analogous to the proofs in [3]: we restrict the representation of the Lie superalgebra G to G_0.

REFERENCES

1. I. N. Bernshteĭn [Joseph N. Bernstein] and D. A. Leĭtes, *A formula for the characters of the irreducible finite-dimensional representations of Lie superalgebras of series Gl and sl*, C. R. Acad. Bulgare Sci. **33** (1980), 1049–1051. (Russian)*

2. N. Bourbaki, *Groupes et algèbres de Lie*, Chap. VIII, Actualités Sci. Indust., no. 1364, Hermann, Paris, 1975.

3. D. A. Leĭtes, *Formulas for the characters of irreducible finite-dimensional representations of simple Lie superalgebras*, Funktsional. Anal. i Prilozhen. **14** (1980), no. 2, 35–38; English transl. in Functional Anal. Appl. **14** (1980).

Translated by L. QUEEN

Editor's note.* For a fuller version of the results of [1], see their paper, *Invariant differential operators and irreducible representations of Lie superalgebras of vector fields*, Serdica **7 (1981), 320–334; English transl., Selecta Math. Sovietica **1** (1981), 143–160 (1982).

Amer. Math. Soc. Transl.
(2) Vol. **132**, 1986

Free Groups, the T-Property, and II_1-Factors with Different Countable Fundamental Groups

UDC 519.4

V. YA. GOLODETS AND N. I. NESSONOV

Abstract. It is proved that there exist II_1-factors with countable fundamental groups which contain any given countable subgroup of $\mathbf{R}_+^*$. It is also proved that the countable free groups F_n with n generators ($n \geq 2$) do not have the T-property and the II_1-factors $R(F_n)$ constructed by the regular representation of F_n are not isomorphic to $R(G)$, where G is a T-group.

There has been recent progress in the study of nonhyperfinite factors and their automorphism groups in the theory of operator algebras (see, for example, [1]–[3]). The T-property of groups was used in [1] for the proof of the existence of II_1-factors with a countable fundamental group, although the problem of generators of this group has remained open.

In this note we shall consider actions of T-groups on injective von Neumann algebras. We shall prove that there exist type II factors with a countable fundamental group containing any given countable subgroup of $\mathbf{R}_+^*$ as a subgroup. Consequently, there exists a continuum of such inequivalent actions. It will also be proved that the countable free groups F_n with n generators ($n \geq 2$) do not have the T-property and the full factors $R(F_n)$ of type II_1, constructed from the regular representation of F_n, are not isomorphic to the full factors $R(G)$, where G is a countable T-group.

1. Let F_n be the free group with n generators $g_1, \dots, g_n$, let $L(F_n)$ be the type II_1 factor induced by the left regular representation of F_n in $l^2(F_n)$, and let λ_g be the unitary operator in $L(F_n)$ corresponding to $g \in F_n$ in the following manner:

$$(\lambda_g f)(h) = f(g^{-1}h) \qquad (f(g) \in e^2(F_n),\ g, h \in F_n).$$

1980 *Mathematics Subject Classification* (1985 *Revision*). Primary 46L35; Secondary 46L40, 20C07.

Translation of Dokl. Akad. Nauk Ukrain. SSR Ser. A **1983**, no. 8, 7–10; MR **85d**:46083.

For $\tilde{\varphi}(\varphi_1, \varphi_2, \dots, \varphi_n) \in \mathbf{T}^n$, where $\mathbf{T}$ is the unit circle in the complex plane, we consider the compact automorphism group $\theta\tilde{\varphi}$ of $L(F_n)$, acting on $\lambda_{g_1}, \dots, \lambda_{g_n}$ according to the formulas

$$\theta_{\tilde{\varphi}}(\lambda_{g_k}) = \varphi_k \lambda_{g_k} \qquad (k = 1, \dots, n).$$

THEOREM 1.1. *If there exists k $(k = 1, \dots, n)$ such that $\varphi_k \neq 1$, then $\theta_{\tilde{\varphi}}$ is an outer automorphism of $L(F_n)$.*

PROOF. If $\theta_{\tilde{\varphi}}$ is inner, then there exists a unitary operator $u = \sum_{g \in F_n} c_g \lambda_g$ for which

$$\left(\sum_{g \in F_n} c_g \lambda_g\right) \lambda_{g_k} = \varphi_k \lambda_{g_k} \left(\sum_{g \in F_n} c_g \lambda_g\right)$$

or $c_{g_k} h g_k^{-1} = \varphi_k c_h$ for all $h \in F_n$ and $k = 1, \dots, n$.

Since F_n is a free group, it is easy to deduce from this that u is a scalar operator.

COROLLARY 1.1. *Let* $\operatorname{Aut} L(F_n)$ *and* $\operatorname{Int} L(F_n)$ *be the groups of automorphisms and of inner automorphisms of $L(F_n)$, respectively. Then, for $n \geq 2$,* $\operatorname{Out} L(F_n) = \operatorname{Aut} L(F_n)/\operatorname{Int} L(F_n)$ *is not discrete in the induced topology.*

COROLLARY 1.2. *The factors $L(G)$, where G is a group with the T-property, are not isomorphic to the factors $L(F_n)$ $(n \geq 2)$.*

Indeed, $\operatorname{Out} L(G)$ is a countable group and $\operatorname{Out} L(F_n)$ is not discrete. Consequently, there exist nonisomorphic full factors.

COROLLARY 1.3. *The groups F_n $(n \geq 2)$ do not have the T-property.*

2. Let M be a factor of type II_∞ and let $\operatorname{Aut} M$ be its group of automorphisms. $\operatorname{Aut} M$ is a separable topological group in the topology defined by the neighborhood base of the identity

$$U_\varphi(\varepsilon) = \{\theta \in \operatorname{Aut} M \colon \|\varphi \circ \theta - \varphi\| < \varepsilon, \varphi \in M_*, \varepsilon > 0\}.$$

Let us denote by $\lambda_1, \dots, \lambda_g$ generators of a countable subgroup $\Gamma \subset \mathbf{R}_+^*$. Let $(R_\Gamma, \varphi) = \bigotimes_i (R_{\lambda_i}, \varphi_i)$, where R_{λ_i} is an approximately finite factor of type III λ_i and φ_i is a faithful normal (f.n.) state on R_{λ_i} whose modular group is periodic with period $T_i = 2\pi/\ln \lambda_i$. Let G be any countable T-group, and $\{M_g\}$ $(g \in G)$ a countable family of factors indexed by the elements of G with $M_g \in R_\Gamma$ for all $g \in G$. Furthermore, let

$$(N, \psi) = \bigotimes_{g \in G} (M_g, \varphi_g).$$

We consider the factor $K = N \times_\alpha G$ which is the crossed product of N and G with respect to the action of α on N induced by the group translation. Let K_ψ be the centralizer of $\tilde{\psi}$ on K constructed from ψ in the standard fashion. It is obvious that $K_\psi = N_\psi \times_\alpha G$. We assume that N_ψ is the GNS-representation of N with respect to ψ, acting in the Hilbert space H. Then K_ψ is generated by the operators $\pi(a)$ and λ_g $(a \in N_\psi, g \in G)$ whose action in $L^2(H, G)$ is defined by

$$\begin{cases} (\pi(a)\xi)(g) = \alpha_g(a)\xi(g), \\ (\lambda_h\xi)(g) = \xi(ph) \qquad (\{\xi(g)\} \in l^2(H,G), a \in N_\psi, g \in G). \end{cases}$$

Let B be a factor of type I_∞, acting in the standard manner in the Hilbert space $L^2(B, \mathrm{tr})$, where tr is a faithful normal semifinite trace on B. Let $M_\Gamma = K_\psi \otimes B$.

THEOREM 2.1. *The subgroup $A_G \subset \operatorname{Aut} M_\Gamma$ generated in the algebraic sense by $\{\theta \in \operatorname{Aut} M_\Gamma \colon \theta(\lambda_g \otimes I) = \lambda_g \otimes 1, g \in G\}$ and $\operatorname{Int} M$ is open and closed in* $\operatorname{Aut} M$.

PROOF. Let J_1 be the antiunitary isometry in K_ψ, sending $x \in L^2(K_\psi, \tilde{\psi})$ to x^*. For any $\theta \in \operatorname{Aut} M_\Gamma$ we can consider the representation

$$\pi_\theta(g) = J_1 \otimes J_2(\lambda_g \otimes I)J_1 \otimes J_2\theta(\lambda_g \otimes I)$$

of G, acting in $L^2(K_\psi, \tilde{\psi}) \otimes L^2(B, \mathrm{tr})$, where J_2 is the antiunitary isometry $J_2 x = x^*$ in $L^2(B, \mathrm{tr})$.

We prove that there exists a neighborhood of the identity in $\operatorname{Aut} M_\Gamma$ belonging to A_G.

Since G has the T-property and the representation

$$\pi_e(g) = J_1 \otimes J_2(\lambda_g \otimes I)J_1 \otimes J_2(\lambda_g \otimes I)$$

leaves invariant the vectors $\xi_0 \otimes \eta \in L^2(K_\psi, \tilde{\psi}) \otimes L^2(B, \mathrm{tr})$, where

$$\xi_0 = (\xi_0(g)) = \begin{cases} 1 \in L^2(N_\psi, \psi) & \text{for } g = e, \\ 0 & \text{for } g \neq e, \end{cases}$$

there is a neighborhood U of the identity in $\operatorname{Aut} M_\Gamma$ with the following property: if $\theta \in U$, then $\pi_\theta(y) = y$ for some $y \in L^2(K_\psi, \tilde{\psi}) \otimes L^2(B, \mathrm{tr})$, $y \neq 0$. Since, according to Segal, y is a closed operator in $L^2(M_\Gamma, \tilde{\psi} \otimes \mathrm{tr})$ affiliated with M_Γ, from $\pi_\theta(y) = y$ we can deduce that

$$\theta(\lambda_g \otimes I)y = y(\lambda_g \otimes I), \qquad y^*\theta(\lambda_g^{-1} \otimes I) = (\lambda_g^{-1} \otimes I)y^* \qquad (g \in G).$$

Therefore

$$y^*y = (\lambda_g^{-1} \otimes I)y^*y(\lambda_g \otimes I) \quad \text{for all } g \in G. \tag{1}$$

Let q be the projection onto the closure of the range of the selfadjoint operator $yy^* \neq 0$. Then $[\lambda_g \otimes I, q] = 0$ $(g \in G)$ (cf. (1)).

Since $R(G)' \cap M_\Gamma = C \otimes B$, where $R(G) = (\lambda_g \otimes I, g \in G)''$, we have $q \in \mathbf{C} \otimes B$. Let $y = u|y|$ be the polar decomposition of y. Taking account of (1), we obtain

$$\theta(\lambda_g \otimes I)u|y| = u|y|(\lambda_g \otimes I) = u(\lambda_g \otimes I)|y|$$

or $\theta(\lambda_g \otimes I)u = u(\lambda_g \otimes I)$ $(g \in G)$.

Set $\tilde{M}_\Gamma = M_\Gamma \otimes B$, $\theta_1 = \theta \otimes \mathrm{id}$, $q_1 = q \otimes I$, and $u_1 = u \otimes I$. Then q_1 and $I - q_1$ are infinite projections in $\tilde{M}_\Gamma$. Therefore, in the unitary group $U(\tilde{M}_\Gamma)$ of $\tilde{M}_\Gamma$ there exists an operator w_1 such that $u = w_1^* q_1$ and

$$\operatorname{Ad} w_1 \circ \theta_1(\lambda_g \otimes I)q_1 = q_1(\lambda_g \otimes I) \qquad (g \in G).$$

Consequently, $\theta_1^{-1} \circ \operatorname{Ad} w_1^*(q_1) \in \mathbf{C} \otimes \tilde{B}$, where $\tilde{B} = B \otimes B$. Since $\theta_1^{-1} \circ \operatorname{Ad} w_1^*(q_1)$ is infinite, we obtain

$$\theta_1 \circ \operatorname{Ad} w_2(\lambda_g \otimes q_1) = \lambda_g \otimes q_1 \qquad (g \in G)$$

for some $w_2 \in U(\tilde{M}_\Gamma)$. On the other hand, $R(G)' \cap \tilde{M}_\Gamma = \mathbf{C} \otimes \tilde{B}$. Therefore,

$$\theta_1 \circ \operatorname{Ad} w_2(I \otimes q_1 \tilde{B} q_1) = I \otimes q_1 \tilde{B} q_1.$$

Consequently, there exists $w_3 \in U(\tilde{M}_\Gamma)$ for which

$$\operatorname{Ad} w_3 \circ \theta_1(K_\psi \otimes q_1) = K_\psi \otimes q_1, \quad \operatorname{Ad} w_3 \circ \theta_1(I \otimes x) = I \otimes x \qquad (x \in q_1 \tilde{B} q_1),$$
$$\operatorname{Ad} w_3 \circ \theta_1(\lambda_g \otimes q_1) = \lambda_g \otimes q_1 \qquad (g \in G). \tag{2}$$

Let $(\tilde{M}_\Gamma)_{q_1}$ be the reduced subalgebra of $\tilde{M}_\Gamma$ corresponding to q_1. Then $\tilde{M}_\Gamma \sim (M_\Gamma)_{q_1} \otimes I_2$. Therefore, by (2) there exists $w_4 \in U(\tilde{M}_\Gamma)$ for which

$$\operatorname{Ad} w_4 \circ \theta_1(K_\psi \otimes I) = K_\psi \otimes I, \quad \operatorname{Ad} w_4 \circ \theta_1(\lambda_g \otimes I) = \lambda_g \otimes I \qquad (g \in G),$$
$$\operatorname{Ad} w_4 \circ \theta_1(I \otimes x) = I \otimes x \qquad (x \in \tilde{B}). \tag{3}$$

Consequently, $\operatorname{Ad} w_4 \circ \theta_1 \in A_G$. Since $\theta_1 = \theta \otimes \mathrm{id}$, we have

$$\operatorname{Ad} w_4(I \otimes I \otimes x) = I \otimes I \otimes x \qquad (x \in B),$$

and therefore $w_4 \in M_\Gamma \otimes I$. Hence $\theta \in A_G$ and $U \subset A_G$. The fact that A_G is closed in $\operatorname{Aut} M_\Gamma$ may be proved by similar (simpler) arguments. This completes the proof of Theorem 2.1.

COROLLARY 2.2. *Let*

$$\operatorname{Aut}_\theta M_\Gamma = \{\theta \in \operatorname{Aut} M_\Gamma \colon (\tilde{\psi} \otimes \mathrm{tr}) \circ \theta = \psi \otimes \mathrm{tr}\}.$$

Then $\operatorname{Aut} M_\Gamma / \operatorname{Aut}_\theta M_\Gamma$ *is countable and discrete.*

Indeed, (3) implies that $A_G \subset \operatorname{Aut}_0 M_\Gamma$, and we only need to use the openness of A_G in $\operatorname{Aut} M_\Gamma$.

COROLLARY 2.3. *The fundamental group of* M_Γ *(and also of* K_ψ*) contains the given countable subgroup* $\Gamma \subset \mathbf{R}_+^*$.

By the construction of $K_\psi \subset K$, if for the projections p_1 and p_2 from K_ψ we have $\psi(p_1)/\psi(p_2) = \lambda_i \in \Gamma$, then K contains a partial isometry u with $u^*u = p_1$ and $uu^* = p_2$; then

$$(K_\psi)_{p_1} \sim (K_\psi)_{p_2} \cdot ((K_\psi)_{p_1} = u^*(K_\psi)_{p_2} u).$$

COROLLARY 2.4. *A* $T(ICC)$*-group has a continuum of inequivalent actions on approximately finite factors of type* II_1.

BIBLIOGRAPHY

1. A. Connes, *A factor of type* II_1 *with countable fundamental group*, J. Operator Theory **4** (1980), 151–153.

2. A. Connes and V. Jones, *A* II_1 *factor with two nonconjugate Cartan subalgebras*, Bull. (N.S.) Amer. Math. Soc. **6** (1982), 211–212.

3. Marie Choda, *Crossed products and property* T, Math. Japon. **26** (1981), 557–567.

Translated by J. SZUCS

Amer. Math. Soc. Transl.
(2) Vol. **132**, 1986

On the Conjugacy of Finite Sets of Subgroups in a Free Product of Groups

I. S. BEZVERKHNYAYA

1. We study the free product $G = A_1 * A_2$ of groups A_1 and A_2. It is well known [1] that each element $w \in G$ can be uniquely written in the canonical form $w = a_1 \cdots a_n$, where all $a_i \neq 1$ and adjacent elements a_i and a_{i+1} belong to distinct factors A_1, A_2. $L(w) = n$ is the *syllable length* of w. The word $a_1 \cdots a_s$ is called an *initial segment* of w, and, if $s = [n/2]$, the *left half* of w. A word w is said to be a *transform* if its left half coincides with the left half of w^{-1}.

Suppose that W is a set of elements of G. We denote by $W_{v,i}$ the set of all transforms in W of the form vKv^{-1}, where $K \in A_i$, $i = 1, 2$. We call the pair (v, i) the *type* of a transform in $W_{v,i}$. We denote the subgroup of G generated by the set $W_{v,i}$ by $A_{v,i} = \mathrm{gp}\{W_{v,i}\}$.

DEFINITION 1 [2]. We call a set of words $W = \{w_i\}_1^N$ in $G = A_1 * A_2$ *Nielsen* if the following four conditions hold:

a) $L(w_{i_1}^{\varepsilon_1} \cdots w_{i_n}^{\varepsilon_n}) \geq L(w_{i_j}^{\varepsilon_j})$, where $\varepsilon = \pm 1$, $w_{i_j} \neq (w_{i_{j+1}}^{\varepsilon_{j+1}})^{-1}$ and it is impossible to find in $w_{i_1}^{\varepsilon_1} \cdots w_{i_n}^{\varepsilon_n}$ a subword $w_{i_j}^{\varepsilon_j} \cdots w_{i_s}^{\varepsilon_s}$, $1 \leq j \leq s \leq n$, in the w_i which is equal to 1 in G.

b) The left half of any nontransform $w_i \in W$ is isolated in W.

c) If $w_\alpha = v_\alpha K_\alpha v_\alpha^{-1}$ has odd length then there is no $w \in \mathrm{gp}\{W \backslash w_\alpha\}$ such that $w_\alpha w = v_\alpha K'_\alpha v_\alpha^{-1}$.

d) If we have two transforms $w_\alpha = v_\alpha K_\alpha v_\alpha^{-1}$ and $w_\beta = v_\beta K_\beta v_\beta^{-1}$, $L(w_\alpha) = L(w_\beta)$, $v_\alpha \neq v_\beta$ and K_α and K_β belong to one of the factors of G, then there is no $w \in \mathrm{gp}\{W\}$ such that $w^{-1}(v_\alpha K_\alpha v_\alpha^{-1})w = v_\beta K_\beta v_\beta^{-1}$.

1980 *Mathematics Subject Classification* (1985 *Revision*). Primary 20E06, 20F10.

Translation of Algorithmic Problems of the Theory of Groups and Semigroups (M. D. Grindlinger [Martin Greendlinger], editor), Tul′sk. Gos. Ped. Inst., Tula, 1981, pp. 102–116; MR **84c**:20032..

Suppose that H is a finitely generated subgroup of G with a Nielsen set of generators W. Each element $h \neq 1$ in H has an irreducible description in W-symbols: $h = w_{i_1}^{\varepsilon_1} \cdots w_{i_s}^{\varepsilon_s}$, $\varepsilon_i = \pm 1$. On multiplying by a sequence of fixed transforms of one type we obtain a description of h in u-symbols: $h = u_1 \cdots u_t$, where $u_i \neq 1$ and $u_i \neq u_{i-1}^{-1}$, $i = 1, \ldots, t-1$.

We call a product of this form a word in the subgroup H. Here

$$u_i \in U = \Big(W \setminus \Big(\bigcup_{(v,i)} W_{v,i}\Big)\Big) \cup \Big(\bigcup (A_{v,i} \setminus \{1\})\Big).$$

It is well known [2], [4] that it is possible to transform an arbitrary finite set W of elements of G by a finite sequence of Nielsen transformations into a Nielsen set where the process is effective if the inclusion problem is solvable for A_1 and A_2.

In what follows we shall consider Nielsen sets, which satisfy Lemma 1.

LEMMA 1. *If $W = \{w_i\}_1^N$ is a Nielsen set then it may be mapped into a Nielsen set $W' = \{w_i'\}_1^N$ such that for any two words*

$$w_i'^{\varepsilon} = l_{1w_i} \cdots l_{kw_i} \cdots l_{nw_i} K_{w_i} r_{nw_i} \cdots r_{1w_i}, \qquad \varepsilon = \pm 1,$$
$$w_j'^{\eta} = l_{1w_j} \cdots l_{kw_j} \cdots l_{mw_j} K_{w_j} r_{mw_j} \cdots r_{1w_j}, \qquad \eta = \pm 1,$$

where $l_{1w_i} \cdots l_{kw_i} \neq l_{1w_j} \cdots l_{kw_j}$, $1 \leq k \leq \min\{m, n\}$, there does not exist a word $w \in \operatorname{gp}\{W'\}$, $L(w) < 2k$, such that

$$w l_{1w_i} \cdots l_{kw_i} = l_{1w_j} \cdots l_{kw_j}.$$

The proof is by induction on word length.

We carry out on the product $h = u_1 \cdots u_t$ all **reductions** in $G = A_1 * A_2$ and obtain a syllable description of h: $h = a_1 \cdots a_n$, $n \geq 1$. Using the properties of Nielsen sets [3] the following lemma can be established.

LEMMA 2. *Each syllable a_i in $h = a_1 \cdots a_n$ has the form $a_i = x_i y_i z_i$, $i = 1, \ldots, n$, where x_i is a central or a right syllable of u_{j-1}, y_i is a central syllable of $u_j \in A_{v,k}$ and z_i is a left or central syllable of u_{j+1}.*

LEMMA 3 (Kurosh lemma for Nielsen sets). *Each transform in $G = A_1 * A_2$ pertaining to a subgroup H generated by a Nielsen set satisfying Lemma 1 has the form*

$$g^{-1} K g = u_1^{-1} \cdots u_n^{-1} u_0 u_n \cdots u_1,$$

where $u_1^{-1} \cdots u_n^{-1} u_0^{-1} u_n \cdots u_1$ is a word in H and u_0 is a transform in $A_{v,i}$.

The proof goes by induction on the length of the word $u_1 \cdots u_k = g^{-1} K g$.

LEMMA 4. *Suppose that a subgroup H of $G = A_1 * A_2$ has the decomposition*

$$H = F * \prod_{i=1}^{m} {}^{*} A_{v_i,1} * \prod_{j=1}^{n} {}^{*} A_{v_j,2} \tag{1}$$

corresponding to its Nielsen generators, and $v_i^{-1}Kv_i$ is a transform in H, where the factor $A_{v_i,1}$ is contained in (1). *Then $v_i^{-1}Kv_i \in A_{v_i,1}$.*

PROOF. By Lemma 3, $v_i^{-1}Kv_i = u_1 \cdots u_n u_0 u_n^{-1} \cdots u_1^{-1}$. Since the right-hand side is a word in H, we can show (see [2]) that it is a simple word. Assuming the contrary, i.e. that $v_i^{-1}Kv_i = u_1 \cdots u_n u_0 u_n^{-1} \cdots u_1^{-1} = w_1 \cdots w_r$ is a product of simple words, on multiplying by w_i^{-1} we could shorten the length of the **wing*** v_1^{-1}, which is impossible since v_i is the left half of a Nielsen transformation. In the simple word $u_1 \cdots u_n u_0 u_n^{-1} \cdots u_1^{-1}$ there may be one or three maximal u-symbols. We show that three is impossible. If there are three then, on applying Lemma 1, it can be shown that they are u_n, u_0, and u_n^{-1}; and $L(u_n) = L(u_0) = L(u_1 \cdots u_n u_0 u_n^{-1} \cdots u_1^{-1})$. Here u_0 is a transform and u_n is a nontransform and has isolated left half. Also, $L(u_1 \cdots u_{n-1}) < L(u_0) = L(u_n)$. Hence by multiplying by the word $u_1 \cdots u_{n-1}$ of shorter length the isolated left half is transformed into the left half of the transform $v_i^{-1}Kv_i$, which is impossible by Lemma 1. Thus the maximal element u_0 is unique. We now show that $u_1 \cdots u_n = 1$ and $L(u_1 \cdots u_n) < L(u_0) = L(u_1 \cdots u_n u_0 u_n^{-1} \cdots u_1^{-1})$.

Cancellation among $u_1 \cdots u_n$ and u_0 does not go as far as the kernel $u_0 = v_i^{-1}K'v_i$, for then it reaches K'. We have therefore that $v_i = u_1^{-1} \cdots u_n^{-1} v_i^{-1}$, and so $u_1 \cdots u_n = 1$. Hence, $u_0 = v_i^{-1}Kv_i$.

LEMMA 5. *Suppose that $G = A_1 * A_2$ and that H_1 and H_2 are subgroups of G with decomposition*

$$H_1 = F_1 * \prod_{i=1}^{m}{}^* A_{v_i,1} * \prod_{j=1}^{n}{}^* A_{v_j,2},$$

$$H_2 = F_2 * \prod_{i=1}^{m'}{}^* A'_{v'_i,1} * \prod_{j=1}^{n}{}^* A'_{v'_j,2},$$

corresponding to their Nielsen generators. If there exists $z \in G$ with $z^{-1}H_1z = H_2$, then $m = m'$, $n = n'$, and there exist $h_i \in H_2$, $i = 1, \dots, m+n$, such that

$$h_i^{-1}z^{-1}A_{v_i,1}zh_i = A'_{v'_i,1} \qquad (1 \le i \le m)$$

and

$$h_{m+j}^{-1}z^{-1}A_{v_{m+j},2}zh_{m+j} = A'_{v'_{m+j},2} \qquad (m+1 \le j \le n).$$

PROOF. Suppose that H_1 is conjugate with H_2: $z^{-1}H_1z = H_2$. We take two distinct transforms $g_1K_1g_1^{-1}$ and $g_1K_2g_1^{-1}$ belonging to some subgroup $z_1^{-1}A_{v_1,1}z$. By hypothesis they are contained in H_2. By Lemma 3 we have

$$g_1K_1g_1^{-1} = u_1 \cdots u_n u_0 u_n^{-1} \cdots u_1^{-1}, \quad g_2Kg_2^{-1} = u'_1 \cdots u'_{n'} u_0 u'^{-1}_{n'} \cdots u'^{-1}_1.$$

*Editor's note. This is a literal translation of the Russian "крыло", introduced in the present context by A. A. Markov (Trudy Mat. Inst. Steklov. **42** (1954), Chapter I, §4). In the English translation (Office of Technical Services, U. S. Dept. of Commerce, Washington, D.C., 1961) the term is rendered as "delimiter".

We shall show that u'_0 and u_0 belong to the same $A'_{v'_i,1}$. We transform the last relations by the subword $u'_1 \cdots u'_{n'}$:

$$\begin{aligned} u''_0 &= u'^{-1}_{n'} \cdots u'^{-1}_1 (g_1 K_1 g_1^{-1}) u'_1 \cdots u'_{n'} \\ &= u'^{-1}_{n'} \cdots u'^{-1}_1 (u_1 \cdots u_n u_0 u_n^{-1} \cdots u_1^{-1}) u'_1 \cdots u'_{n'} \\ u'_0 &= u'^{-1}_{n'} \cdots u'^{-1}_1 (g_2 K_2 g_2^{-1}) u'_1 \cdots u'_{n'}. \end{aligned}$$

Since $u'_0 \in A'_{v'_j,1}$ we have $u''_0 \in A_{v_j,2}$ by Lemma 4. We shall prove that u''_0 and u_1 are transforms of the same type. Let us assume that $L(u_0) > L(u''_0) = L(u'_0)$.

We reduce the right-hand side of the expression

$$u''_0 = u'^{-1}_{n'} \cdots u'^{-1}_1 (u_1 \cdots u_n u_0 u_n^{-1} \cdots u_1^{-1}) u'_1 \cdots u'_{n'} \tag{2}$$

to a word. It will contain a transform of type u_0. Hence, by property 1 of Nielsen sets it follows that the length of this word in G is greater than or equal to that of u_0. We have reached a contradiction.

In a similar way it is possible to prove that one cannot have $L(u_0) < L(u'_0)$. Hence $L(u_0) = L(u'_0) = L(u''_0)$. It follows from this that (2) is a simple word.

If $u''_0 = v_j K_1 v_j^{-1}$ and $u_0 = v_i K_2 v_i^{-1}$, and $v_i \neq v_j$, that is, u_0 and u''_0 are transforms of different type, then it follows from (2) that transforms of different type can be transformed one into the other, which is impossible. Thus u_0 and u''_0 are transforms of the same type. Hence u'_0 and u''_0 are also of the same type. We show that the words $u_1 \cdots u_n$ and $u'_1 \cdots u'_n$ in (2) are equal. For this we reduce the product $u'^{-1}_{n'} \cdots u'^{-1}_1 u_1 \cdots u_n$ to a word in H. If this product is 1 then we are home. Suppose not. Then $u'^{-1}_{n'} \cdots u'^{-1}_1 u_1 \cdots u_n = u''_1 \cdots u''_k$. We substitute this in (2):

$$u''_0 = u''_1 \cdots u''_k u_0 u''^{-1}_k \cdots u''^{-1}_1. \tag{3}$$

If we suppose that (3) is not a word, it follows that u''_k is a transform of type u_0. Combining the transforms of one type $u''_k u_0 u''^{-1}_k = \tilde{u}_0$, we get $u''_0 = u''_1 \cdots u''_{k-1} \tilde{u}_0 u''^{-1}_{k-1} \cdots u''^{-1}_1$. This expression is a word, and u''_0 and $\tilde{u}_0$ are of the same type. Also

$$L(u''_0) = L(u''_1 \cdots u''_{k-1} \tilde{u}_0 u''^{-1}_{k-1} \cdots u''^{-1}_1) = L(\tilde{u}_0);$$

that is, $u''_1 \cdots u''_{k-1} \tilde{u}_0 u''^{-1}_{k-1} \cdots u''^{-1}_1$ is a simple word. By arguing as in Lemma 4 one can show that $u''_1 \cdots u''_{k-1} = 1$. Then $u'^{-1}_{n'} \cdots u'^{-1}_1 u_1 \cdots u_n = u''_k$. Hence $u_1 \cdots u_n = u'_1 \cdots u'_{n'} u''_k$. On the left there is a word, and on the right also since $u'_1 \cdots u'_{n'} u'_0 u'^{-1}_{n'} \cdots u'^{-1}_1$ is a word and $u'_{n'}$ and u'_0 are not transforms of the same type and so neither are $u'_{n'}$ and u''_k. Therefore $n = n' + 1$, and $u_1 = u'_1, \dots, u_n = u''_k$. However this contradicts the fact that $u'_1 \cdots u_n u_0 u_n^{-1} \cdots u_1^{-1}$ is a word; that is, $u''_k = 1$. By the same token there exists $h_i \in H_2$ such that

$$h_i^{-1} z^{-1} A_{v_i,j} z h_i \subseteq A'_{v_i,j}, \qquad j = 1, 2.$$

By inverting the expression $H_1 = z^{-1} H_2 z$ it can be shown that the converse inclusion holds, and so

$$h_i^{-1} z^{-1} A_{v_i,j} z h_i = A'_{v_i,j}.$$

2. We consider an algorithm for constructing the normalizer of a subgroup of a free product.

DEFINITION 2. If the normalizer of each finitely generated subgroup H of a group G is finitely generated then we shall say that G *has property* R.

LEMMA 6. *Suppose that* $G = A_1 * A_2$ *and* $H = B_1 * B_2$, *where* $B_1 \leq A_1$ *and* $B_2 \leq A_2$. *Then* H *is self normalizing:* $N_G(H) = H$.

The proof is obvious.

THEOREM 1. *If in* $G = A_1 * A_2$ *both* A_1 *and* A_2 *have property* R, *then so does* G.

PROOF. We assume that the finitely generated subgroup H of G has a free product decomposition corresponding to a Nielsen set of generators:

$$H = F * \prod_{i=1}^{m} {}^{*} A_{v_i,1} * \prod_{j=1}^{n} {}^{*} A_{v_{m+j},2}. \tag{4}$$

There are two possibilities:

Case 1. The decomposition (4) contains a factor conjugate with a subgroup of a factor of G. We find all elements $z \in N_G(H)$ which generate $N_G(H)$. Our problem is to prove that there is a finite set of such elements if H is to be finitely generated. $z^{-1}Hz = H$ and

$$z^{-1}Hz = z^{-1}(F * A_{v_1,1} * \cdots * A_{v_m,1} * A_{v_{m+1},2} * \cdots * A_{v_n,2})z.$$

Here $A_{v_1,1} = \{v_1^{-1}Kv_1 \mid K \in A_1\}$. We shall conjugate this equation by v_1:

$$v_1Hv_1^{-1} = v_1z^{-1}Hzv_1^{-1} = v_1z^{-1}\left(F * \prod_{i=1}^{m} {}^{*} A_{v_i,1} * \prod_{j=1}^{n} {}^{*} A_{v_j,2}\right) zv_1^{-1}$$

$$= v_1F_1v_1^{-1} * v_1A_{v_i,1}v_1^{-1} * \cdots * v_1A_{v_{m+n},2}v_1^{-1}, \qquad D = v_1A_{v_1}v_1^{-1}.$$

Here $D = v_1A_{v_1,1}v_1^{-1}$ is a subgroup of G generated by the kernels of the generators of $A_{v_1,1}$; and $D \leq A_1$. We reduce the generators of $v_1Fv_1^{-1} * D * \cdots * v_1A_{v_{m+n},2}v_1^{-1}$ to a Nielsen set. In consequence they may be written thus:

$$F * D * \cdots * A'_{v'_{m+n},2} \tag{5}$$

The subgroup D is unaffected by reducing the set of generators to a Nielsen set. We assume that among the generators of length 1 there will be reduced generators of some $v_iA_{v_i,1}v_i^{-1}$, $i \neq 1$. Combining this with D we reduce the number of factors in (5), which is impossible by Lemma 5. Nontransforms cannot go over into transforms since the rank of the free subgroup F would be more than that of F_1. Thus we have

$$v_1z^{-1}(F * A_{v_1,1} * \cdots * A_{v_i,1} * \cdots)zv_1^{-1} = F_1 * D * \cdots * A'_{v'_{m+n},2} = H_2 \tag{6}$$

By Lemma 5 there exist i and $h_i \in H_2$ such that $h_iv_iz^{-1}A_{v_i,1}zv_1^{-1}h_i^{-1} = D \leq A_1$. The kernels of $A_{v_1,1}$ belong to A_1. Moreover, $A_{v_i,1} = v_i^{-1}C_iv_i$, $C_i \leq A_1$,

$h_i v_1 z^{-1} v_i^{-1} C_i v_i z v_1^{-1} h_i^{-1} = D$, and $h_i v_1 z^{-1} v_i^{-1} = a_0 \in A_1$ (see [1], p. 188). We find the form of a_0. We transform

$$a_0(v_i F v_i^{-1} * v_i A_{v_i,1} v_i^{-1} * \cdots * C_i * \cdots) a_0^{-1} = F * D * \cdots * A'_{v'_{m+n},2}. \quad (7)$$

We choose an arbitrary generator x from the subgroup $v_i F v_i^{-1} * v_i A_{v_i,1} v_i^{-1} * \cdots * C_i * \cdots$ of length greater than 1 (if there are no such elements then the subgroup contains one or two factors and the proof reduces to using Lemma 6). Therefore suppose that $x = a_1 b_1 a_2 b_2 \cdots$. We transform it by a_0: $a_0 x a_0^{-1} = u_1 \cdots u_k \in H_2$.

Suppose that $u_1 \in D$. Then

$$a_0(a_1 b_1 a_2 b_2 \cdots) a_0^{-1} = \tilde{a}_0 a'_1 b'_1 \cdots;$$

also $u_1 = \tilde{a}_0$, $u_2 = a'_1 b_1 \cdots$, and $a_0 a_1 = \tilde{a}_0 a'_1$. Hence $a_0 = a_0 a'_1 a_1^{-1}$; but

$$a_0 = h_i v_1 z^{-1} v_i^{-1} = \tilde{a}_0 a'_1 a_1^{-1}. \quad (8)$$

Since $\tilde{a}_0 \in D$, $\tilde{a}_0 \in H_2$. We conjugate (7) by a_0:

$$\tilde{a}_0^{-1} h_i v_1 z^{-1} v_i^{-1} (v_i H v_i^{-1}) v_i z v_1^{-1} h_i^{-1} \tilde{a}_0 = H_2 = v_1 H v_1^{-1}.$$

Moreover,

$$v_1^{-1} \tilde{a}_0^{-1} h_i v_1 z^{-1} H z v_1^{-1} h_i^{-1} \tilde{a}_0 v_1 = H, \qquad v_1^{-1} \tilde{a}_0 v_1 \cdot v_1^{-1} h_i v_1 \in H.$$

We now obtain z using (8):

$$a_0 = h_i v_1 z^{-1} v_i^{-1} = \tilde{a}_0 a'_1 a_1^{-1}; \qquad v_1^{-1}(h_i v_1 z^{-1} v_i^{-1}) = v_1^{-1} \tilde{a}_0 a'_1 a_1^{-1};$$

$$(v_1^{-1} h_i v_1) z^{-1} = v_1^{-1} \tilde{a}_0 a'_1 a_1^{-1} v_i; \qquad (v_1^{-1} h_i v_1) z^{-1} = (v_1^{-1} \tilde{a}_0 v_1)(v_1^{-1} a'_1 a_1^{-1} v_i);$$

$$z^{-1} = (v_1^{-1} h_i v_i)^{-1} (v_1^{-1} \tilde{a}_0 v_1)(v_1^{-1} a'_1 a_1^{-1} v_i).$$

Thus the adjunction of z to the normalizer of H is equivalent to adjoining to it the element $v_1^{-1} a'_1 a_1^{-1} v_i$. But only a finite number of such elements can be constructed. Hence a_1 is the first syllable of some generating subgroup $v_i F v_i^{-1} * v_i A_{v_i,1} v_i^{-1} * \cdots * C_i * \cdots$. Now a'_1 is the initial syllable of some Nielsen generator of H_2. But these subgroups have a finite number of generators which begin or end with syllables from A_1, and so the set of such words is finite. There is also only a finite set of elements v_i, v_1 since there is only a finite number of transforms in A_1.

Case 2. The decomposition (4) contains no factor conjugate with a subgroup of a factor of G. In this case $H = F = \langle w_1, \ldots, w_k \rangle$. If $u_1, u_2 \in F$ and $z \in N_G(F)$, then $z_0 = u_1 z u_2 \in N_G(F)$. We shall therefore choose z to be shortest in the double coset FzF so that for arbitrary $u_1, u_2 \in F$ we must have $L(zu_1) \geq L(z)$, $L(u_2 z) \geq L(z)$, and $L(u_2 z u_1) \geq L(z)$. Among the generators $\{w_i\}$ we select those which are cyclically irreducible. If such do not occur we choose $w_1 = cxc^{-1}$, where x is a cyclically irreducible element. We pass from F to $c^{-1}Fc$, which has the cyclically irreducible generator x. We reduce the set of generators of F_1 to a Nielsen set. If subgroups are conjugate, so are their normalizers. Let

$$L_0 = \max\{L(w_1), \ldots, L(w_n)\},$$

where the w_i are Nielsen generators of F_1. We show that the length of the element z as chosen does not exceed one half of L_0: $L(z) \leq [\frac{1}{2}L_0]$.

We consider the product $z^{-1}xz = u_1 \cdots u_n$. Suppose that there is no cancellation between x and z. If cancellation does occur between z^{-1} and x then less than half of x is involved in the cancellation, since otherwise the length of z can be reduced, which is impossible by the minimality of z. We assume that $L(z) > [\frac{1}{2}L_0]$. We represent the product $u_1 \cdots u_n$ in the form of a product of simple words:

$$z^{-1}xz = u_1 \cdots u_n = v_1 \cdots v_p,$$

where $L(v_p) \leq L_0$. Consider $v_{p-1}v_p$. Here cancellation does not affect the kernels of v_{p-1} or v_p. Transforming $v_1 \cdots v_p$ by v_p, one can shorten the length of z:

$$v_p z^{-1} x z v_p^{-1} = v_p v_1 \cdots v_{p-1}$$

(more than one half of the length of v_p is in z). Then $p > 1$, which is impossible.

Suppose that $p = 1$: $z^{-1}xz = v_1$. Then $L(z) > \frac{1}{2}L_0$, and $L(z^{-1}x) > \frac{1}{2}L_0$, which is impossible. Thus $L(z) \leq [\frac{1}{2}L_0]$.

Using the structure of simple words, one can prove that z can be taken to be a subword of the left or right half of some Nielsen generator of F_1. After this the proof of the finite generation of the normalizer is straightforward.

It follows from the proof of Theorem 1 that Theorem 2 holds.

THEOREM 2. *Suppose that $G = A_1 * A_2$. If the factors A_i $(i = 1, 2)$ have property R and if in each of them the problems of inclusion and construction of the normalizer of a finitely generated subgroup are solvable, then there exists an algorithm to construct the normalizer $N_G(H)$ of any finitely generated subgroup H of G.*

3. We now consider the coset intersection problem for a free product of groups.

DEFINITION 3. We shall say that the coset intersection problem for cosets of a finite number of finitely generated subgroups $B_1, \ldots, B_N$ of a group B is *solvable* if there is an algorithm to determine for arbitrary $x_1, \ldots, x_N \in B$ whether or not $x_1B_1 \cap \cdots \cap x_NB_N$ is empty.

Let $H_1, \ldots, H_N$ be a finite set of finitely generated subgroups of a group G.

THEOREM 3. *If the intersection problem for a finite number of cosets of finitely generated subgroups $H_1, \ldots, H_N$ is solvable in each factor of $G = A_1 * A_2$, then it is solvable in G.*

We remark that the solution of the coset intersection problem implies the solution of the inclusion problem for the group. It therefore follows from the hypothesis of the theorem that the inclusion problem is solvable in the factors A_1 and A_2 of $G = A_1 * A_2$. Hence a Nielsen set of generating subgroups H_j $(j = 1, \ldots, N)$ can be constructed [2], [4]. We shall use the method of types which was introduced in [3].

We denote the Nielsen set of generating subgroups H_j by $W^{(j)} = \{w_i\}_1^{n_j}$, $j = 1, \dots, n$.

We shall assume that these Nielsen sets satisfy Lemma 1.

We therefore need to establish whether or not $s_1H_1 \cap \dots \cap s_NH_N$ is empty, where $s_i \in G$, $i = 1, \dots, N$. In the general case $s_i \notin H_i$: if $s_i \in H_i$ we may put $s_i = 1$ (the coset coincides with H_i). Clearly, we may equally well consider the intersection

$$s_N^{-1}s_1H_1 \cap \dots \cap s_N^{-1}s_{N-1}H_{n-1} \cap H_N.$$

We denote $s_N^{-1}s_i = y_i$. If (9) is not empty, then there exist $h_i \in H_i$ such that $y_1h_1 = \dots = y_{N-1}h_{N-1} = y_N$. We write each h_i in terms of u-symbols from the corresponding subgroup:

$$\begin{aligned} v &= y_1u^{(1)}(1)u^{(1)}(2)\cdots u^{(1)}(r_1) = y_2u^{(2)}(1)\cdots y_2u^{(2)}(r_2) \\ &= \dots = y_{N-1}u^{(N-1)}(1)\cdots u^{(N-1)}(r_{N-1}) = u^{(N)}(1)\cdots u^{(N)}(r) \\ &= y_0q(1)q(2)\cdots q(r). \end{aligned} \tag{10}$$

The last line gives an irreducible expression for v in G, where y_0 is an initial subword of v and $L(y_0) = \max_{(i)} L(y_i)$. We put $R^{(i)} = W^{(i)} \cup \overline{W}^{(i)} \cup \{\alpha_i\}$, $i = 1, \dots, N$, where $x \in \overline{W}^{(i)}$ if $x^{-1} \in W^{(i)}$ and α_i is an arbitrary symbol. We also denote by T^i the set of all types of transforms contained in $W^{(i)}$ augmented by the symbol ρ_i. The relations (10) define the mapping

$$\begin{aligned} \psi_v(1, 2, \dots, r) &= ((R^{(1)} \times \mathbf{Z}) \times T^1 \times (R^{(1)} \times \mathbf{Z})) \\ &\quad \times ((R^{(2)} \times \mathbf{Z}) \times T^2 \times (R^{(2)} \times \mathbf{Z})) \\ &\quad \times \dots \times ((R^{(N)} \times \mathbf{Z}) \times T^N \times (R^{(N)} \times \mathbf{Z})) \\ &= \prod_{i=1}^{N} ((R^{(i)} \times \mathbf{Z}) \times T \times (R^{(i)} \times \mathbf{Z})), \end{aligned} \tag{11}$$

where $\mathbf{Z} = \{0, 1, 2, \dots\}$.

By Lemma 2 each syllable in the last line of (10) is made up of elements with three adjacent symbols; that is,

$$q(j) = x^{(1)}y^{(1)}z^{(1)} = \dots = x^{(N)}y^{(N)}z^{(N)}.$$

Then the mapping (11) defines the function

$$\begin{aligned} \psi_v(j) = (&((r_1^{(1)}(j), p_1^{(1)}(j)), \tau^{(1)}(j), (r_2^{(1)}(j), p_2^{(1)}(j))), \\ &((r_1^{(2)}(j), p_1^{(2)}(j)), \tau^{(2)}(j), (r_2^{(2)}(j), p_2^{(2)}(j))), \dots, \\ &((r_1^{(N)}(j), p_1^{(N)}(j)), \tau^{(N)}(j), (r_2^{(N)}(j), p_2^{(N)}(j)))), \end{aligned}$$

where $r_1^{(i)}(j) = \alpha^{(i)}$ and $p_1^{(i)}(j) = 0$ if $x^{(i)} = 1$, and $r_1^{(i)}(j) = w_1^{(i)}(j) \in R^{(i)} \backslash \{\alpha_i\}$ if $x^{(i)}$ is the $p_1^{(i)}(j)$th syllable of $w_1^{(i)}(t)$; $\tau^{(i)}(j) = \rho^{(i)}$ if $y^{(i)} = 1$, and $\tau^{(i)}(j) = (v^{(i)}, k^{(i)})$ if $y^{(i)}$ is a syllable of a transform of type $(v^{(i)}, k^{(i)})$ ($k^{(i)} = 1, 2$); $\tau_2^{(i)}(j) = \alpha^{(i)}$ if $z^{(i)} = 1$, and $r_2^{(i)}(j) = w_2^{(i)}(t)$ if $z^{(i)}$ is the $p_2^{(i)}(j)$th syllable of

$$w_2^{(i)}(t) \in R^{(i)} \backslash \{\alpha_i\}, \ i = 1, \dots, N.$$

In order to continue with the proof of Theorem 3 we need Definition 4 and Lemma 7.

DEFINITION 4. We shall say that $\psi_v(j)$ is the *type* of the syllable $q(j)$ in the word v.

The cardinality of the set of all types does not exceed the number

$$\mathcal{A} = \prod_{i=1}^{N}\left(2\sum_{w^{(i)}\in W^{(i)}} L(w^{(i)}) - \mathcal{B}^{(i)} + 1\right)^2 (\sigma^{(i)} + 1),$$

where $\mathcal{B}^{(i)}$ is the number of transforms in $R^{(i)}\backslash\{\alpha_i\} = [W^{(i)} \cup \overline{W}^{(i)}]$, and $\sigma^{(i)}$ is the number of types of transforms in $W^{(i)}$, $i = 1, \dots, N$.

LEMMA 7. *If $L(v) > L_0 + \mathcal{A}$, where v is the word in* (10) *and $v \in H_N$, then there is a word $v' \in H_N$, $L(v') < L(v)$, which satisfies* (10).

The proof is analogous to that of Lemma 1 of [6].

We proceed with the proof of Theorem 3.

We define a mapping of the Nielsen set $W^{(i)}$ of generators of the subgroups H_i thus:

$$\Phi^{(i)}(w_j)\begin{cases} w_j & \text{if } w_j \text{ is a nontransform in } W^{(i)},\\ v_t^{(i)-1}\mu_t^{(i)}v_t^{(i)} & \text{if } w_j \in A^{(i)}_{v_t^{(i)},1},\ 1 \le t \le m^{(i)},\\ v^{(i)}_{t+m^{(i)}}\mu^{(i)-1}_{t+m^{(i)}}v^{(i)-1}_{t+m^{(i)}} & \text{if } w_j \in A^{(i)}_{v^{(i)}_{t+m^{(i)}},2},\ 1 \le t \le n^{(i)},\end{cases}$$

where $A^{(i)}_{v_i^{(i)},1}$ and $A^{(i)}_{v^{(i)}_{t+m},2}$ are free factors of a decomposition of H_i corresponding to the Nielsen set $W^{(i)}$, then

$$H_i = F^{(i)} * \prod_{i=1}^{m^{(i)}} {}^* A^{(i)}_{v_t^{(i)},1} * \prod_{j=1}^{n^{(i)}} {}^* A^{(i)}_{v^{(i)}_{m^{(i)}+j},2},$$

where $\mu_t^{(i)}$ is some symbol.

We extend the mapping to H_i. Suppose that $w = u_1^{(i)} \cdots u_k^{(i)}$ is a word written in u-symbols. Then

$$\Phi^{(i)}(w) = \Phi^{(i)}(u_1^{(i)}) \cdots \Phi^{(i)}(u_k^{(i)}).$$

We shall assume that for arbitrary $j, 1 \le j \le m^{(i)}$, and $a \in A^{(i)}_{j,1}$

$$\Phi^{(i)}(v_j^{(i)-1}av_j^{(i)}) = v_j^{(i)-1}\mu_j^{(i)}v_j^{(i)}),$$

and for arbitrary j, $1 \le j \le n^{(i)}$, and $b \in A^{(i)}_{m^{(i)}+j,2}$

$$\Phi^{(i)}(v^{(i)-1}_{m^{(i)}+j}bv^{(i)}_{m^{(i)}+j}) = v^{(i)-1}_{m^{(i)}+j}\mu^{(i)}_{m^{(i)}+j}v^{(i)}_{m^{(i)}+j}.$$

Here

$$A^{(i)}_{j,1} = v_j^{(i)}A^{(i)}_{v_j^{(i)},1}v_j^{(i)-1}, \qquad A^{(i)}_{m^{(i)}+j,2} = v^{(i)}_{m^{(i)}+j}A^{(i)}_{v^{(i)}_{m^{(i)}+j},2}v^{(i)-1}_{m^{(i)}+j}.$$

The words $y_1, y_2, \dots$ from (10) are known. Our problem is to find words $u^{(i)}(1)\cdots u^{(i)}(r_i) \in H_i$, $i = 1, \dots, N$, satisfying (10). By Lemma 7,

$$L(u^{(N)}(1)\cdots u^{(N)}(r_N)) \le L_0 + \mathcal{A} = K',$$

$$L(u^{(i)}(1)\cdots u^{(i)}(r_i) \le 2L_0 + \mathcal{A} = K \qquad (i \ne N).$$

In each H_i we consider the set of all words $\Phi^{(i)}(u^{(i)}(1))\cdots\Phi^{(i)}(u^i(r_i))$ whose preimages satisfy the condition

$$L(u^{(i)}(1)\cdots u^{(i)}(r_i)) < K$$

(or K', $i = N$). The set $H_\Phi^{(i)}$ of these words (images) is finite, and since a preimage can be written in the form $u_1 \cdots u_k = (x_1 y_1 z_1)\cdots(x_k y_k z_k)$, the subgroup generated by the kernels $A_{j(i),n}^{(i)}$ $(n = 1, 2)$ corresponds in the image to the single element $\mu_{j(i)}^{(i)}$. This eliminates the possibility of an infinite set of kernels y_i, and the number of elements x_i, z_i is finite for words of bounded length.

From all of the sets $H_\Phi^{(i)}$ we choose an arbitrary word and consider its product with the corresponding y_i:

$$\begin{aligned}
&y_1\Phi^{(1)}(u^{(1)}(1)) \cdot \Phi^{(1)}(u^{(1)}(2)) \cdots \Phi^{(1)}(u^{(1)}(r_1)) \\
&= y_1(x_1^{(1)}\mu_{i_1}^{(1)}z_1^{(1)})(x_z^{(1)}\mu_{i_2}^{(1)}z_2^{(1)})\cdots(x_{r_1}^{(1)}\mu_{i_{r_1}}^{(1)}z_{r_1}^{(1)}) \\
&= y_2\Phi^{(2)}(u^{(2)}(1)) \cdot \Phi^{(2)}(u^{(2)}(2)) \cdots \Phi^{(2)}(u^{(2)}(r_2)) \\
&= y_2(x_1^{(2)}\mu_{i_2}^{(2)}z_1^{(2)})(x_2^{(2)}\mu_{i_2}^{(2)}z_2^{(2)})\cdots(x_{r_2}^{(2)}\mu_{i_{r_2}}^{(2)}z_{r_2}^{(2)}) \\
&= \cdots = \Phi^{(N)}(u^{(N)}(1)) \cdot \Phi^{(N)}(u^{(N)}(2)) \cdots \Phi^{(N)}(u^{(N)}(r_N)) \\
&= (x_1^{(N)}\mu_{i_1}^{(N)}z_1^{(N)})(x_2^{(N)}\mu_{i_2}^{(N)}z_2^{(N)})\cdots(x_{r_N}^{(N)}\mu_{i_{r_N}}^{(N)}z_{r_N}^{(N)}).
\end{aligned} \tag{12}$$

In the first $N-1$ lines conjugacy is possible between y_i and $\Phi^{(i)}(u_i(1)\cdots)$. We consider the first line. $y_1 = y_{1n}ay_{1n}$. Suppose that a subword $\Phi^{(1)}$ of length j_1 occurs in the reduction of y_1. If y_{1n} ends in the syllable g and $\Phi^{(1)}$ begins with the syllable $x_1^{(1)}\mu_{i_1}^{(1)}z_1^{(1)}$ and there is cancellation between them, then g and the preimage of $x_1^{(1)}\mu_1^{(1)}z_1^{(1)}$ are contained in one and the same factor A_i $(i = 1, 2)$. Moreover some factor $A_{v_{i_1}^{(1)},1}^{(1)}$ of H_i corresponds to $\mu_{i_1}^{(1)}$. Since there is cancellation, $g^{-1} = x_1^{(1)}\mu_{i_1}^{(1)}z_1^{(1)}$, whence

$$\mu_{i_1}^{(1)} = x_1^{(1)-1}g^{-1}z_1^{(1)-1}. \tag{13}$$

Cancellation is possible if there is in $A_{v_{i_1}^{(1)},1}^{(1)}$ a transform with kernel (13). This can be clarified since the inclusion problem is solvable for the factors A_i $(i = 1, 2)$. If $x_1^{(1)-1}g^{-1}z_1^{(1)-1} \in A_{v_{i_1}^{(1)},2}^{(1)}$, then we proceed to find the kernel of the next syllable of the formal word. If there are no such transforms, cancellation is impossible. Cancellations are carried out in all the equations (12). Next, among the syllables of the formal equations corresponding to relations (10) we can set up a one-to-one correspondence. Using it we define the $\mu_{j(i)}^{(i)}$ not already found. Thus we shall get the relations

$$x_{j(1)}^{(1)}\mu_{j(1)}^{(1)}z_{j(1)}^{(1)} = \cdots = x_{j(N)}^{(N)}\mu_{j(N)}^{(N)}z_{j(N)}^{(N)}.$$

However, if one $\mu_{j(i)}^{(i)}$ here is known then the rest are defined at once. We consider the case where none of the $\mu_{j(i)}^{(i)}$ are known. If $N = 2$ then the intersection problem for cosets of two arbitrary subgroups is easily solved, and we have

$$x_{j(1)}^{(1)}\mu_{j(1)}^{(1)}z_{j(1)}^{(1)} = x_{j(2)}^{(2)}\mu_{j(2)}^{(2)}z_{j(2)}^{(2)}$$

and

$$x_{j(1)}^{(1)}z_{j(1)}^{(1)}(z_{j(1)}^{(1)-1}\mu_{j(1)}^{(1)}z_{j(1)}^{(1)}) = x_{j(2)}^{(2)}z_{j(2)}^{(2)}(z_{j(2)}^{(2)-1}\mu_{j(2)}^{(2)}z_{j(2)}^{2}).$$

Hence

$$(z_{j(2)}^{(2)-1}x_{j(2)}^{(2)-1}x_{j(1)}^{(1)}z_{j(1)}^{(1)})(z_{j(1)}^{(1)-1}\mu_{j(1)}^{(1)}z_{j(1)}^{(1)}) = z_{j(2)}^{(2)-1}\mu_{j(2)}^{(2)}z_{j(2)}^{(2)}. \tag{15}$$

Suppose that $\mu_{j(1)}^{(1)}$ corresponds to $A_{j(1),1}^{(1)}$ and $\mu_{j(2)}^{(2)}$ to $A_{j(2),1}^{2}$. Equation (15) amounts to solving the question of whether or not the intersection of cosets

$$(z_{j(2)}^{(2)-1}x_{j(2)}^{(2)-1}x_{j(1)}^{(1)}z_{j(1)}^{(1)})(z_{j(1)}^{(1)-1}A_{j(1),1}^{(1)}z_{j(1)}^{(1)}) \cap z_{j(2)}^{(2)-1}A_{j(2),1}^{(2)}z_{j(2)}^{(2)}$$

is empty. By hypothesis there is an algorithm solving the coset intersection problem in the factors of the free product. Hence there is also an algorithmic solution to the intersection problem for cosets of two subgroups of a free product. The problem for N groups has an analogous solution. Equation (14) corresponds to the intersection of cosets in a single factor of G.

4. The following theorem is fundamental.

THEOREM 4. *Suppose that the factors in $G = A_1 * A_2$ have property R. Suppose that in A_1 and A_2 the following conditions hold:*

1) *The coset intersection problem for finitely generated subgroups is solvable.*
2) *The problem of constructing a normalizer is solvable.*
3) *There exists an algorithm which establishes the conjugacy of two sets of subgroups.*

Then there is an algorithm to find for two sets of finitely generated subgroups $\{H_i\}$, $\{H_i'\}$ whether there exists $z \in G$ such that $\&_1^N (z^{-1}H_iz = H_i')$.

PROOF. By [5] it is possible to effectively decide whether or not two subgroups H_i and H_i' are conjugate in a free product. In our case the solution of the inclusion problem follows from that of the coset intersection problem, and conjugacy is given by hypothesis, so that the hypotheses of a theorem of Moldavanskiĭ [5] are satisfied. We assume that for the sets $\{H_i\}_1^N$ and $\{H_i'\}_1^N$ there is a $z \in G$ such that

$$\mathop{\&}_{i=1}^{N} (z^{-1}H_iz = H_i'), \tag{16}$$

so that to every subgroup H_i there corresponds a conjugate subgroup H_i'. For the pairs (H_i, H_i') we can define $z_i \in G$ such that $z_i^{-1} H_i z_i = H_i'$, $i = 1, \ldots, N$. From this and (16) follow the relations

$$\mathop{\&}_{i=1}^{N} (z z_i^{-1} H_i z_i z^{-1} = H_i);$$

that is, $z z_i^{-1} \in N_G(H_i)$. Hence we have

$$\mathop{\&}_{i=1}^{N} (z_i z^{-1} = h_i),$$

where $h_i \in N_G(H_i)$. Therefore

$$z = z_1 h_1^{-1} = z_2 h_2^{-1} = \cdots = z_N h_N^{-1}. \tag{17}$$

It follows from this that finding z satisfying (16) is equivalent to solving (17). This solution is the intersection of cosets of finitely generated subgroups. From Theorem 3 and the hypothesis it follows that (17) has an effective solution. The theorem is proved.

DEFINITION 5. We say that a group B *has property* $\mathcal{H}$ (or *satisfies* $\mathcal{H}$) if the intersection of finitely generated subgroups is finitely generated.

LEMMA 8. *If a group B 1) satisfies $\mathcal{H}$, 2) has solvable coset intersection problem for any two finitely generated subgroups, and 3) has an algorithm for writing down the intersection of any two finitely generated subgroups, then the coset intersection problem for an arbitrary finite number of cosets of finitely generated subgroups is solvable in B.*

From Theorem 4 and Lemma 8 there follows

THEOREM 4′. *Suppose that $G = A_1 * A_2$, where A_i $(i = 1, 2)$ satisfies R and $\mathcal{H}$. Then if in the factors 1) the coset intersection problem is solvable for cosets of two finitely generated subgroups, 2) there exists an algorithm to find generators of the intersection of two finitely generated subgroups, 3) the problem of constructing a normalizer is solvable, and 4) there is an algorithm to determine the conjugacy of two sets of subgroups, then there is an algorithm in G to determine whether, for any two sets of subgroups, there is a $z \in G$ such that*

$$\mathop{\&}_{i=1}^{N} (z^{-1} H_i z = H_i').$$

BIBLIOGRAPHY

1. Wilhelm Magnus, Abraham Karrass, and Donald Solitar, *Combinatorial group theory*, Interscience, 1966.

2. V. N. Bezverkhniĭ, *Nielsen's reduction method for a free product of groups*, Collection of Scientific Works of Dept. of Higher Math., Tul'sk. Politekhn. Inst., Tula, 1972, pp. 44–70. (Russian) R. Zh. Mat. **1973**, 9A218.

3. V. N. Bezverkhniĭ and È. V. Rollov, *Subgroups of a free product of groups*, Modern Algebra (E. S. Lyapin, editor), Vyp. 1, Leningrad. Gos. Ped. Inst., Leningrad, 1974, pp. 16–31. (Russian)

4. D. I. Moldavanskiĭ, *Nielsen's method for a free product of groups*, Ivanov. Gos. Ped. Inst. Uchen. Zap. **61** (1969), 170–182. (Russian)

5. ——, *Conjugacy of subgroups of a free product*, Ivanov. Gos. Ped. Inst. Uchen. Zap. **106** (1972), 123–135. (Russian) R. Zh. Mat. **1973**, 1A204.

6. I. S. Bezverkhnyaya, *An analogue of Whitehead's result for a finite set of subgroups of a free group and for a free product of groups*, Applied Mathematics, Tul'sk. Politekhn. Inst., Tula, 1976, pp. 48–55. (Russian) R. Zh. Mat. **1977**, 11A246.

Translated by J. C. LENNOX

Amer. Math. Soc. Transl.
(2) Vol. **132**, 1986

On Normalizers of Artin Groups

V. A. GRINBLAT

An Artin group is a group, call it G, given on generators a_i, $i \in I$, with defining relations $a_i a_j a_i \cdots = a_j a_i a_j \cdots, i, j \in I$, where the words standing on the left and right consist of m_{ij} letters, with a_i and a_j alternating; here the m_{ij} are the entries of a Coxeter matrix $M = (m_{ij})$, $i, j \in I$, of type 1 (see [1]). On adding the relations $a_i^2 = 1$, $i \in I$, to the defining relations of G, we get a presentation of a Coxeter group $\overline{G}$. If $\overline{G}$ is finite, the corresponding group G is said to be an Artin group of finite type. Since Coxeter introduced them in 1935, Coxeter groups have been extensively investigated; an exposition of the results about them can be found in Bourbaki [1]. Of the Artin groups, only the braid groups have been studied so far, except of course for free groups; the most important achievement of recent years has been the solution of the conjugacy problem, to be found in Garside [2] and Makanin [3].

Artin groups have been studied by Brieskorn and Saito [4], the methods being combinatorial and very similar to those of Garside. The word and conjugacy problems for an Artin group G with finite Coxeter group $\overline{G}$ were solved in [4], and a description of the center was given.

In §1 we consider the centralizers of arbitrary elements of an Artin group G of finite type. We prove that, for each element A of G, the centralizer N_A of A in G is generated by finitely many elements of G^+, and we give an algorithm that constructs generators for N_A, for arbitrary A. This problem was solved for the braid groups in [5].

1980 *Mathematics Subject Classification* (1985 *Revision*). Primary 20F36; Secondary 20F10.

Translation of Algorithmic Problems of the Theory of Groups and Semigroups (M. D. Grindlinger [Martin Greendlinger], editor), Tul′sk. Gos. Ped. Inst., Tula, 1981, pp. 82–94; MR **84b**: 20047.

In §2 we prove that in Artin groups (semigroups) of finite type, the intersection problem for cyclic subgroups (subsemigroups) generated by arbitrary elements W_1 and W_2 such that $W_1W_2 = W_2W_1$ is solvable.

§1. Words in the alphabet a_1, a_1^{-1}, a_2, $a_2^{-1}, \ldots$ will be called words in the group G. Words in the alphabet $a_1, a_2, \ldots$ will be called positive words. The empty word is denoted by Λ, the length of a word W by $L(W)$ and $W \doteq V$ means letter-for-letter equality of the words W and V. If $W = a_{i_1}^{\varepsilon_1} \cdots a_{i_v}^{\varepsilon_v}$ is a word in the alphabet of G, where $\varepsilon_r = \pm 1$, then the word $a_{i_v}^{-\varepsilon_v} \cdots a_{i_1}^{-\varepsilon_1}$ is called the word inverse to W and is denoted by W^{-1}.

An Artin semigroup is the semigroup G^+ generated by generators a_i with defining relations $a_i a_j a_i \cdots = a_j a_i a_j \cdots$. Since the defining relations for G^+ are homogeneous, it follows that words which are equal in G^+ have equal lengths, and therefore that the word problem in G^+ is solvable. We denote equality of words in G^+ by $\doteq$.

With every Coxeter matrix M we associate the concepts of left and right divisibility in the semigroup G_M^+. If U and V are positive words, then every positive word W positively equivalent to the word UV is by definition divisible on the left by U and on the right by V with respect to M. By a common multiple of a system of elements g_j, $j \in J$, of G^+ we mean an element of G^+ divisible on the left by each of the elements g_j. *A* least common multiple is a common multiple that divides every other common multiple of the system g_j.

The following assertion is proved in [4] by using a division algorithm developed there.

PROPOSITION 1. *A finite set of elements of an Artin semigroup G^+ either has a least common multiple or has no common multiple.*

DEFINITION 1 [4]. Let M be a Coxeter matrix of type 1, and let $J \subset I$ be a subset such that the letters in J have a common multiple in G_M^+. Then the uniquely determined least common multiple Δ_J of the letters in J in G_M^+ is called the *fundamental element* of J in G_M^+.

LEMMA 1 [4]. *Let $J = \{j_1, \cdots, j_k\}$ be a finite subset of I such that the fundamental element Δ_J exists in G_M^+. Then*

$$\Delta_J = [a_{j_1} \cdots a_{j_k}]_l [a_{j_1} \cdots a_{j_k}]_r,$$

where $[a_{j_1} \cdots a_{j_k}]_l$ and $[a_{j_1} \cdots a_{j_k}]_r$ denote the left and right least common multiples of $a_{j_1}, \ldots, a_{j_k}$ respectively.

LEMMA 2 [4]. *If $J \subset I$ is a subset having fundamental element Δ_J, then there is a unique involutory automorphism σ_J of the semigroup G_J^+ having the following properties*:

(i) *σ_J takes letters to letters; that is, $\sigma_J(a_j) = a_{\sigma(j)}$ $\forall j \in J$, where σ is a permutation of J such that $\sigma^2 = \mathrm{id}$ and $m_{\sigma(i),\sigma(j)} = m_{ij}$ $\forall i, j \in J$.*

(ii)

$$W\Delta_J = \Delta_J \sigma_J(W) \quad \textit{for all } W \in G_J^+. \tag{1}$$

It is clear that the following is an immediate consequence of (1):

$$W^{\varepsilon}\Delta^{\beta} = \Delta^{\beta}\sigma(W^{\varepsilon}). \tag{2}$$

The first assertion of the following lemma follows immediately from (1):

LEMMA 3 [4]. *Suppose that the fundamental element Δ_J exists. Then the following assertions hold for arbitrary U, V, and W.*

(i) *Δ_J divides U on the left if and only if it divides U on the right.*

(ii) *If Δ_J divides the product VW, then every letter a_j, $j \in J$, either divides V on the right or divides W on the left.*

If a positive word has the form $W = UbbV$, where U and V are positive words and b is a letter, we shall say that the word W contains a square. We say that a positive word W is *squarefree* if there is no word V positively equivalent to W that contains a square.

The following lemma establishes important properties of the fundamental element.

LEMMA 4 [4]. *If the subset $J \subset I$ has fundamental element Δ_J, then the following assertions are true:*

(i) *An element $U \in G_J^+$ is squarefree if and only if it divides Δ_J.*

(ii) *The least common multiple of a set of squarefree elements of G_J^+ is squarefree.*

The following lemma was proved in [4] in a slightly different formulation.

LEMMA 5. *If the element $U \in G^+$ is squarefree and the letter a_j does not divide U on the left (right), then the word a_jU (Ua_j respectively) is squarefree.*

An Artin semigroup G^+ having a fundamental element can be characterized in terms of properties of the canonical map onto the Artin group G.

PROPOSITION 2. *The following assertions are equivalent for every Coxeter matrix M.*

(i) *There is a fundamental element Δ in G^+.*

(ii) *Every finite subset of G^+ has a least common multiple.*

(iii) *The canonical map $G^+ \to G$ is injective, and for every $A \in G$ there exist $B, C \in G^+$ such that $A = BC^{-1}$ and the image of C in G lies in the center.*

THEOREM [4]. *Let M be a Coxeter matrix. A fundamental element Δ exists in G^+ if and only if the group $\overline{G}$ is finite.*

ARTIN'S THEOREM [4]. *The word problem is solvable in Artin groups of finite type.*

GARSIDE'S THEOREM [4]. *If the positive words V and W are equal (conjugate) in the group G, then they are equal (conjugate) in the semigroup G^+.*

LEMMA 6. *If $UV = \Delta$ in G^+, then $V\sigma(U) \doteq \sigma(V)U \doteq \Delta$.*

PROOF. In G we have $\sigma(V)U = \sigma(V)UVV^{-1} = \sigma(V)\Delta V^{-1}$. Use Lemma 2:

$$\sigma(V)U = \sigma(V)\Delta V^{-1} = \Delta\sigma^2(V)V^{-1} = \Delta VV^{-1} = \Delta.$$

Since $\sigma(V)U$ and Δ are positive words, Garside's theorem gives that $\sigma(V)U = \Delta$ in G^+.

LEMMA 7. *Let H_1 be a squarefree word such that $AH_1 = Ca_i$. If every letter a_p dividing A on the right divides H_1 on the left, then a_i divides H_1 on the right.*

PROOF. We argue by contradiction. Suppose that a_i does not divide H_1 on the right. Then by Lemma 5, the word H_1a_i is squarefree. For H_1a_i there is a squarefree word H_2 such that $H_1a_i\sigma(H_2) \doteq \Delta$.

Consider the word $D \doteq AH_1H_2\sigma(H_1) = Ca_iH_2\sigma(H_1) = C\Delta$. By Lemma 3, every letter a_j, $j = 1, \ldots, n$, either divides A on the right or divides $H_1H_2\sigma(H_1)$ on the left. The set of letters dividing $H_1H_2\sigma(H_1)$ on the left contains the set of letters dividing H_1 on the left, and by assumption this last contains the set of letters dividing A on the right. Consequently every letter a_j, $j = 1, \ldots, n$, divides $H_1H_2\sigma(H_1)$ on the left. Thus there exists a word B such that $H_1H_2\sigma(H_1) = B\Delta$. Since $\Delta \doteq a_iH_2\sigma(H_1)$, it follows that $H_1H_2\sigma(H_1) \doteq Ba_iH_2\sigma(H_1)$. Canceling in this last equation, we get that $H_1 \doteq Ba_i$. We have arrived at a contradiction.

LEMMA 8 [5]. *If H_1 is a squarefree word and $AH_1B \doteq C\Delta$, then there are squarefree words H_2 and H_3 such that $A \doteq A_1H_2$, $B \doteq H_3B_1$, and $H_2H_1H_3 \doteq \Delta$.*

PROOF. With every word H_1 we associate the parameter $q(H_1) = L(\Delta) - L(H_1)$, and carry out the proof by induction on $q(H_1)$. If $q(H_1) = 0$, then $L(H_1) = L(\Delta)$, so that $H \doteq \Delta$. Since the empty word is squarefree, the lemma is clear in this case.

Suppose that the lemma holds for any H such that $q(H) \leq k$, $k \geq 0$. We establish it for H with $q(H) = k + 1$. Three cases are possible.

Case I. Suppose that $A \doteq Da_p$, where a_p does not divide H_1 on the left. By assumption $DA_pH_1B = C\Delta$; and by Lemma 5, a_pH_1 is squarefree. Since $q(a_pH_1) = k$, by the inductive hypothesis there exist squarefree words F_1 and F_2 such that $D \doteq A_1F_1$, $B \doteq F_2B_1$, and $F_1a_pH_1F_2 \doteq \Delta$. Therefore a_pH_1 is squarefree.

Case II. Suppose that $B = a_pQ$, where a_p does not divide H_1 on the right. The arguments here are similar to those in Case I.

Case III. The set of letters dividing B on the left is contained in the set of letters dividing H_1 on the right, and the set of letters dividing A on the right is contained in the set of letters dividing H_1 on the left.

Since the set of letters dividing B on the left is contained in the set of letters dividing AH_1 on the right, it is clear from Lemma 3 that every letter a_j, $j = 1, \cdots, n$, divides AH_1 on the right. This means that $AH_1 \doteq E\Delta$ for some E. Since the set of letters dividing A on the right is contained in the set of letters dividing H_1 on the left, it follows from Lemma 3 that H_1 is divisible on the left by all letters a_j, $j = 1, \cdots, n$. Thus $H_1 = \Delta$ and $q(H_1) = 0$, whereas $q(H_1) = k + 1$. This contradiction proves the lemma.

The following lemma, proved for braid groups in [5], can easily be carried over to Artin groups of finite type.

LEMMA 8′. *If* $AX \doteq XB$ *in* G^+, *then* $X = H_1 \cdots H_k$, *where every* H_j *is squarefree, and there exists a sequence* $A_0, A_1, \dots, A_k$ *of words such that* $A_0 \equiv A$, $A_k \equiv B$, *and* $A_{i-1}H_i \doteq H_iA_i$ $(i = 1, 2, \dots, k)$.

Following [5], we shall call a sequence of positive words

$$A_0, H_1, A_1, H_2, A_2, \dots, A_{k-1}, H_k, A_k, \tag{3}$$

where $k > 0$, a *basic sequence* for the word A_0, if

1) every H_i is a nonempty squarefree word,
2) $A_{i-1}H_i \doteq H_iA_i$ $\forall i = 1, 2, \dots, k$;
3) if $0 \leq i \leq k-1$ and $0 \leq j \leq k-1$, then $A_i \neq A_j$ in G^+, and
4) there exists $v \in \{1, \dots, k\}$ such that $A_k \doteq A_{v-1}$.

Two basic sequences, (3) and

$$B_0, G_1, B_1, G_2, \dots, B_{m-1}, G_m, B_m \tag{4}$$

will be regarded as the same if 1) $m = k$, 2) $A_i \doteq B_i$, $i = 1, \dots, k$, and 3) $G_i \doteq H_i$ $\forall i = 1, \dots, k$. Otherwise the two sequences are regarded as different.

LEMMA 9 [5]. *If* $L(A_0) = p$ *in the basic sequence* (3), *then* $k \leq n^p$.

For, since G^+ is a homogeneous semigroup and $L(A_i) = p$ $\forall i = 1, \dots, k$, there exist n^p different words of length p in the alphabet $a_1, \dots, a_n$. This means that $k \leq n^p$.

LEMMA 10 [5]. *For every word* A_0 *of length* p, *there exist only finitely many different basic sequences for* A_0.

PROOF. We prove that a word A_0 of length p has not more than $(n+2)^\alpha$ different basic sequences, where

$$\alpha = n^p L(\Delta) + (n^p + 1)p + 2n^p + 1. \tag{5}$$

If $L(A_0) = p$ in the basic sequence (3), the length of the whole sequence cannot exceed $kL(\Delta) + (k+1)p + 2k + 1$, where $kL(\Delta)$ is the maximum possible length of the word $H_1 \cdots H_k$, $(k+1)p$ is the length of the word $A_0, A_1, \dots, A_k$, and $2k+1$ is the number of commas. Since $k \leq n^p$, the length of a basic sequence for A_0 is not greater than α. Since different basic sequences for such a word cannot be letter-for-letter equal, A_0 cannot have more than $(n+2)^\alpha$ different basic sequences.

LEMMA 11 [5]. *There exists an algorithm which describes all basic sequences of an arbitrary positive word* A_0.

Following [5], we shall say that the basic sequence (3) *generates the basic word*

$$W \equiv H_1H_2 \cdots H_{v-1}H_v \cdots H_kH_{v-1}^{-1} \cdots H_2^{-1}H_1^{-1}. \tag{6}$$

Together with the definition of basic sequence, this definition means that the basic word is uniquely defined by the basic sequence.

We say that W is a basic word associated with the word A if there exists a basic sequence for A that generates W.

Basic words relating to one and the same word A are said to be the same if they are equal in G, and different if they are different in G.

LEMMA 12 [5]. *If $L(A_0) = p$, then there exist only finitely many basic words relating to A_0.*

PROOF. We show that the number of basic words relating to A_0 is not more than $(n+1)^\nu$, where $\nu = 2n^p L(\Delta)$. For, if $\partial(A_0) = p$ in the basic sequence (3), then the length of the basic word generated by it does not exceed $2kL(\Delta)$, since $L(H_j) \le L(\Delta)\ \forall j$, and $k \le n^p$ by Lemma 9. Thus the length of the basic word is not more than $\nu = 2n^p L(\Delta)$, and there are not more than $(n+1)^\nu$ basic words relating to A_0.

From Lemma 11 and Artin's theorem we at once get

LEMMA 13 [5]. *There exists an algorithm describing the different basic words relating to an arbitrary word A_0.*

LEMMA 14 [5]. *Let A be an arbitrary positive word, and $W_1, \dots, W_t$ a listing of all different basic words relating to A. Then among $W_1, \dots, W_t$ there is a word equal to Δ^2 in G^+, or a word equal to Δ in G^+.*

The following two lemmas are analogues of lemmas from [5].

LEMMA 15. *If W is a basic word relating to A_0, then $A_0 W = W A_0$ in the group G.*

LEMMA 16. *If $A_0 X = X A_0$ in G^+, then $X = W_1 \cdots W_r$, $r \ge 0$, in G, where each W_j is a basic word relating to A_0.*

From the lemmas in [4] there follows easily:

LEMMA 17. *For every word Z in G there exist a positive word X and an integer $v \ge 0$ such that $Z = \Delta^{-2v} X$.*

LEMMA 18 [5]. *Let A be a positive word such that $AZ \doteq ZA$. Then $Z = W_1^{\varepsilon_1} \cdots W_p^{\varepsilon_p}$, $p \ge 0$, $\varepsilon_i = \pm 1$, and each W_i is a basic word relating to A.*

THEOREM 1. *There exists an algorithm which constructs, for an arbitrary element C of an Artin group G of finite type, elements $K_1, \dots, K_w$ such that each K_i permutes with C, and every element V permuting with C lies in the subgroup of G generated by $K_1, \dots, K_w$.*

The proof of this theorem goes just as in [5].

§**2.** We consider now the intersection problem for cyclic subsemigroups of Artin semigroups of finite type, and the analogous problem for cyclic subgroups of Artin groups of finite type.

We shall say that two cyclic subgroups (subsemigroups) $\langle W_1\rangle$ and $\langle W_2\rangle$ *have an intersection* if there exist $m, n \in \mathbf{Z}\backslash\{0\}$ ($m, n \in \mathbf{N}$ respectively) such that $W_1^m = W_2^n$.

We shall say that the intersection problem for cyclic subgroups (subsemigroups) is solvable in the group G (in the semigroup G^+) if there exists an algorithm which determines, for any two cyclic subgroups (subsemigroups), whether their intersection is trivial or nontrivial.

LEMMA 19. *Let W_1, $W_2 \in G^+$, where G^+ is an Artin semigroup. Then the equation $W_1^n = W_2^m$ has a solution if and only if there is a solution of the equation $V_1^k = V_2^k$, where $V_1 = W_1^{L(W_2)}$ and $V_2 = W_2^{L(W_1)}$.*

The proof follows easily from the homogeneity of an Artin semigroup.

THEOREM 2. *Suppose that W_1, $W_2 \in G^+$ and $W_1W_2 = W_2W_1$, where G^+ is an Artin semigroup of finite type. Then the intersection problem for the cyclic semigroups $\langle W_1\rangle$ and $\langle W_2\rangle$ is solvable.*

PROOF. Using Lemma 19, we can go over from the words W_1 and W_2 to $V_1 = W_1^{L(W_2)}$ and $V_2 = W_2^{L(W_1)}$; we consider the equation

$$V_1^k = V_2^k. \tag{7}$$

It is clear that V_1 and V_2 are both delta words (for the definition see [6]), or both not delta words.

Case 1. V_1 and V_2 are not delta words; since $V_1V_2 = V_2V_1$, in G we have

$$(V_1V_2^{-1})^k = 1. \tag{8}$$

We rewrite $V_1V_2^{-1} = H_{j_s}\cdots H_{j_1}\Delta^p$ (see [6]), with p even and $p \in \mathbf{Z}$, $H_{j_s}\cdots H_{j_1} \in G^+$. For (8) to hold it is necessary that $(H_{j_s}\cdots H_{j_1})^k \in \Delta^2$; that is, $H_{j_s}\cdots H_{j_1}$ must be a delta word. We consider all possible conjugate powers of $H_{j_s}\cdots H_{j_1}$ with irreducible bases [6]. We shall establish in a finite number of steps either that $H_{j_s}\cdots H_{j_1}$ is a delta word or that we can find a conjugate power with irreducible basis having length greater than $2m_0^{m_0} + 2m_0$. (Then by Lemma 6 of [6], $H_{j_s}\cdots H_{j_1}$ is not a delta word, and so (8) does not hold.) Suppose that $H_{j_s}\cdots H_{j_1}$ is a delta word; then there exist a number n_0 and an even L such that

$$(H_{j_s}\cdots H_{j_1})^{n_0} = \Delta^L. \tag{9}$$

Substituting $V_1V_2^{-1} = H_{j_s}\cdots H_{j_1}\Delta^p$ in (8) and writing $k = n_0t$, we get

$$(H_{j_s}\cdots H_{j_1}\Delta^p)^k = (H_{j_s}\cdots H_{j_1})^{n_0t}\Delta^{pn_0t} = \Delta^{Lt+pn_0t} = 1.$$

Hence

$$t(L + pn_0) = 0, \qquad L + pn_0 = 0. \tag{10}$$

Case 2. V_1 and V_2 are delta words.

LEMMA 20. *If V_1 and V_2 are delta words (that is, $\exists n_0$, n_0' and even L, L' such that $V_1^{n_0} = \Delta^L$ and $V_2^{n_0'} = \Delta^{L'}$), then equation (7) has a solution if and only if $Ln_0' = L'n_0$.*

PROOF. Suppose that (7) holds for some k. Then $(V_1^k)^{n_0 n_0'} = (V_2^k)^{n_0 n_0'}$, so that $\Delta^{Ln_0'k} = \Delta^{L'n_0k}$, and thus $Ln_0' = L'n_0$.

THEOREM 3. *Let W_1 and W_2 be arbitrary words in an Artin group G of finite type such that $W_1W_2 = W_2W_1$. Then the intersection problem for the cyclic subgroups $\langle W_1\rangle$ and $\langle W_2\rangle$ of G is solvable.*

The proof of the theorem is equivalent to solving the equation

$$W_1^n = W_2^m, \tag{11}$$

where $m, n \in \mathbf{Z}$. The solution of (11) is equivalent to that of the disjunction

$$W_1^n = W_2^m \quad \text{or} \quad (W_1^{-1})^n = W_2^m, \tag{12}$$

where $n, m \in \mathfrak{N}$. We shall consider an equation of the form $W_1^n = W_2^m$, $n, m \in \mathfrak{N}$. We write the words W_1 and W_2 in the form

$$W_1 = H_{j_s} \cdots H_{j_1} \delta^{p_1}, \qquad W_2 = H_{i_t} \cdots H_{i_1} \Delta^{p_2},$$

where $p_1, p_2 \in \mathbf{Z}$, both even, and $H_{j_s} \cdots H_{j_1}$ and $H_{i_t} \cdots H_{i_1}$ both lie in G^+. Suppose that $n \geq m$. We set $y_0 = \max\{|p_1|, |p_2|\}$, and transform equation (12) into the following:

$$\begin{gathered} \Delta^{-2y_0n}\Delta^{2y_0n}W_1^n = \Delta^{-2y_0m}\Delta^{2y_0m}W_2^m, \\ (\Delta^{2y_0}W_1)^n = \Delta^{2y_0(n-m)}(\Delta^{2y_0}W_2)^m. \end{gathered} \tag{13}$$

We set $\Delta^{2y_0}W_1 = \tilde{W}_1$ and $\Delta^{2y_0}W_2 = \tilde{W}_2$, $\tilde{W}_1, \tilde{W}_2 \in G^+$. Then (13) takes the form

$$\Delta^{2y_0(n-m)}\tilde{W}_2^m = \tilde{W}_1^n. \tag{14}$$

Since $n \geq m$, we have an equation in G^+. Suppose that $L(\tilde{W}_2) = k_2L_0 + r_2$ and $L(\tilde{W}_1) = k_1L_0 + r_2$, where $L_0 = L(\Delta)$. From (14) we get a condition on length:

$$2y_0(n-m)L_0 + (k_2L_0 + r_2)m = (k_1L_0 + r_1)n.$$

Then $(2y_0L_0 - k_1L_0 - r_1)n = (2y_0n_0 - k_2L_0 - r_2)m$. Two cases are possible:

Case 1. $2y_0n_0 - k_1L_0 - r_1 = 0$, and $2y_0n_0 - k_2L_0 - r_2 \neq 0$. Then $m = 0$, and thus equation (14) makes sense when considered for $\tilde{W}_2 = 1$; that is, when W_2 is a power of Δ. In this case (14) has the form $\Delta^{2y_0n} = \tilde{W}_1^n$. Thus the problem reduces to elucidating whether or not $\tilde{W}_1$ is a delta word.

Case 2. $2y_0L_0 - k_1L_0 - r_1 \neq 0$ and $2y_0L_0 - k_2L_0 - r_2 \neq 0$. Then

$$n = \frac{2y_0L_0 - k_2L_0 - r_2}{2y_0L_0 - k_1L_0 - r_1}m = cm, \quad \text{where } c = \frac{2y_0L_0 - k_2L_0 - r_2}{2y_0L_0 - k_1L_0 - r_1}.$$

Since $n \geq m$, $c \geq 1$. If $c = 1$, then $n = m$ and (14) has the form

$$\tilde{W}_2^n = \tilde{W}_1^m, \tag{15}$$

and the problem reduces to the problem for semigroups considered above. We note that if (15) has a solution, so does (11); if (15) has no solution, neither does (11). If $c > 1$, then setting $n = cm$ in (14), we get

$$\Delta^{2y_0(c-1)m}\tilde{W}_2^m = \tilde{W}_1^{cm}. \tag{16}$$

If $c \in \mathfrak{N}$, we rewrite (16) in the form

$$(\Delta^{2y_0(c-1)}\tilde{W}_2)^m = (\tilde{W}_1^c)^m \tag{17}$$

Once again, (11) reduces to the solution of (17) in the semigroup G^+. If $c \in R$, that is, $c = q/p$, $c > 1$, then $n = (q/p)m$ and $n = qt$, $m = pt$. Substituting n and m in (14), we get

$$\Delta^{2y_0(qt-pt)}\tilde{W}_2^{pt} = \tilde{W}_1^{nt}$$

Thus

$$(\tilde{W}_2^p\Delta^{2y_0(q-p)})^t = (\tilde{W}_1^q)^t. \tag{18}$$

Since the words $\tilde{W}_2^p\Delta^{2y_0(q-p)}$ and $\tilde{W}_1^{n_0}$ are positive, the solution of (14) reduces to that of (18) in the semigroup G^+.

Case 3. $2y_0L_0 - k_1L_0 - r_1 = 2y_0L_0 - k_2L_0 - r_2 = 0$. Then

$$k_1L_0 + r_1 = k_2L_0 + r_2 = 2y_0L_0, \quad \text{i.e. } L(\tilde{W}_1)L(\tilde{W}_2);$$

since $W_1W_2 = \tilde{W}_2\tilde{W}_1$, we can rewrite (14) in the form

$$\Delta^{2y_0(n-m)}(\tilde{W}^{-1})^{n-m} = (\tilde{W}_1\tilde{W}_2^{-1})^m. \tag{19}$$

LEMMA 21. *Suppose that $\tilde{W}_1$ is a positive word. Then $\tilde{W}_1$ is a delta word if and only if there exists $n \in \mathfrak{N}$ such that $(W_1^{-1})^n \in \langle\Delta^2\rangle$.*

PROOF. If $\tilde{W}_1$ is a delta word, then $\exists n \in \mathfrak{N}$ and $L \in \mathbf{Z}$, L even, such that $\tilde{W}_1^n = \Delta^L$. Then $(W_1^n)^{-1} = \Delta^{-L}$; that is, $(\tilde{W}_1^{-1})^n \in \langle\Delta^2\rangle$. If $\tilde{W}_1$ is not a delta word, and $(\tilde{W}_1^{-1})^k \in \langle\Delta^2\rangle$, then $W_1^k \in \langle\Delta^2\rangle$, which means that $\tilde{W}_1^k = \Delta^L$, where $L \in \mathfrak{N}$, L even.

Suppose that the $\tilde{W}_1$ in (19) is not a delta word. Then it follows from Lemma 21 that there is no $n \in \mathfrak{N}$ such that $(W_1^{-1})^n \in \langle\Delta^2\rangle$. Consequently, there is no $n \in \mathfrak{N}$ such that $(\tilde{W}_1\tilde{W}_2^{-1})^n \in \langle\Delta^2\rangle$. By Garside's theorem, $\tilde{W}_1^{-1} = W_1^*\Delta^{\bar{p}_1}$ and $\tilde{W}_1\tilde{W}_2^{-1} = W_2^*\Delta^{\bar{p}_2}$, where $W_1^*, W_2^* \in G^+$, $\bar{p}_1, \bar{p}_2 \in \mathbf{Z}$ are even, and W_1^* and W_2^* are not delta words. Suppose that $y_0 > |p_1|$ in (13). Then (19) takes the form

$$(W_1^*\Delta^{2y_0+\bar{p}_1})^{n-m} = (W_2^*\Delta^{\bar{p}_2})^m = W_2^{*m}\Delta^{m\bar{p}_2}$$

We set $V_1 \rightleftharpoons W_1^*\Delta^{2y_0+\bar{p}_1}$. Then

$$V_1^{n-m} = W_2^{*m}\Delta^{m\bar{p}_2}. \tag{20}$$

We write down a condition on length for equation (20). Suppose that $L(V_1) = L_0k_1 + r_1$ ($L(V_1) \neq 0$) and $L(W_2^*) = L_0k_2 + r_2$, where $L_0 = L(\Delta)$. Then

$$(n-m)(L_0k_1 + r_1) = (L_0k_2 + r_2)m + m\bar{p}_2L_0,$$

$$n(L_0k_1 + r_1) = (L_0k_2 + r_2 + \bar{p}_2L_0 + r_1 + L_0k_1)m.$$

We consider the case where $L_0k_2 + r_2 + \bar{p}_2L_0 + r_1 + L_0k_1 \neq 0$. Then $n = cm$, where $c \geq 1$ and $c \in R$. If $c = 1$, then $n = m$, and the problem reduces to a solution of the analogous problem in the semigroup. If $c > 1$ and $c \in \mathfrak{N}$, then (20) has the form $V_1^{m(c-1)} = (W_2^*)^m\Delta^{m\bar{p}_2}$; if $\bar{p}_2$ is negative, we have

$$(V_1^{(c-1)}\Delta^{-\bar{p}_2})^m = (W_2^*)^m. \tag{21}$$

The problem of solving (14) has thus been reduced to the solution of (21) in G^+. If (21) has a nontrivial solution in G^+, then the m solving it and the corresponding n will be a solution to (14). Conversely, if m, n is a solution of (14) for the given c ($n = cm$), then m is a solution to (21). If $c > 1$, and $c = q/p$ is a fraction in lowest terms, then as $n = qm/p$, we have $m = pt$ and $n = qt$, and equation (21) transforms to (18).

Suppose that $\tilde{W}_1$ in (19) is a delta word. Then $\tilde{W}_2$ also is a delta word. We represent the words W_1 and W_2 in (11) in the form $W_1 = \overline{W}_1\Delta^{p_1}$ and $W_2 = \overline{W}_2\Delta^{p_2}$, where $p_1, p_2 \in \mathbf{Z}$, both even, and $\overline{W}_1, \overline{W}_2 \in G^+$. Since $\tilde{W}_1$ and $\tilde{W}_2$ are delta words, so are $\overline{W}_1$ and $\overline{W}_2$. Thus there exist $n_0, n_0' \in \mathfrak{N}$ and even $L, L' \in \mathfrak{N}$ such that

$$\overline{W}_1^{n_0} = \Delta^L, \qquad \overline{W}_2^{n_0'} = \Delta^{L'}$$

Raising both sides of (11) to the n_0n_0'th power, we get

$$(W_1^{n_0n_0'})^n = (W_2^{n_0n_0'})^m, \qquad (\overline{W}_1^{n_0n_0'}\Delta^{p_1n_0n_0'})^n = (\overline{W}_2^{n_0n_0'}\Delta^{p_2n_0n_0'})^m.$$

Thus

$$(\Delta^{Ln_0'+p_1n_0n_0'})^n = (\Delta^{L'n_0+p_2n_0n_0'})^m.$$

It makes sense to consider the following cases:

a) Suppose that $L' + p_2n_0' \neq 0$ and $L + p_1n_0 \neq 0$. In this case the solution of the equation is obvious, since the problem reduces to that of the intersection of two delta subgroups, $\langle\Delta^{c_1}\rangle$ and $\langle\Delta^{c_2}\rangle$.

b) $L_1' + p_2n_0' = L + p_1n_0 = 0$. Then $W_1^{n_0n_0'} = 1$ and $W_2^{n_0n_0'} = 1$. The problem reduces to that of the intersection of two cyclic subgroups of finite order. This completes the proof of the theorem.

REFERENCES

1. N. Bourbaki, *Groupes et algèbres de Lie*, Chaps. 4–6, Actualités Sci. Indust., no. 1337, Hermann, Paris, 1968.

2. F. A. Garside, *The braid group and other groups*, Quart. J. Math. Oxford Ser (2)**20** (1969), 235–254.

3. A. S. Makanin, *The conjugacy problem in braid groups*, Dokl. Akad. Nauk SSSR **182** (1968), 495–496; English transl. in Soviet Math. Dokl. **9** (1968).

4. Egbert Brieskorn and Kyoji Saito, *Artin-Gruppen und Coxeter-Gruppen*, Invent. Math. **17** (1972), 245–271.

5. G. S. Makanin, *Normalizers in the braid group*, Mat. Sb. **86(128)** (1971), 171–179; English transl. in Math. USSR Sb. **15** (1971).

6. V. N. Bezverkhniĭ and V. A. Grinblat, *The root problem in Artin groups*, Algorithmic Problems in the Theory of Groups and Semigroups (M. D. Grindlinger [Martin Greendlinger], editor), Tul′sk. Gos. Ped. Inst., Tula, 1981, pp. 72–81. (Russian)

Translated by J. WIEGOLD

Amer. Math. Soc. Transl.
(2) Vol. 132, 1986

Groups of Class 2 and Exponent p with Commutator Subgroup of Order p^2

UDC 512.542.3

A. L. VISHNEVETSKIĬ

Suppose that G is a finite p-group of class 2 and that G' is the commutator subgroup of G. The problem of finding defining relations for G in the case $|G'| = p$ was solved in [1] and [2], but it is very difficult for $|G'| \geq p^2$ (see [3]). We solve the problem here for $|G'| = p^2$ and $\exp G = p$.

Thus we suppose that $\exp G = p$ and $|G'| = p^2$. Since G is nonabelian, $p \neq 2$. We associate with G a pair of skew-symmetric matrices over the field F of residue classes $\bmod p$ in the usual way [3]. The following result follows from [4], Theorem 3.

THEOREM 1. *$G = \langle H_1, H_2\rangle$, where H_1 and H_2 are abelian groups and $H_1 \cap H_2 \subset Z$, the center of G.*

We introduce the following pencils of matrices:

$$1)\quad L_\varepsilon = \left.\begin{pmatrix} \lambda & \mu & & & \\ & \lambda & \mu & & \\ & & \ddots & \ddots & \\ & & & \lambda & \mu \end{pmatrix}\right\} \varepsilon;$$

$$L'_\eta = \left.\begin{pmatrix} \lambda & & & \\ \mu & \lambda & & \\ & \mu & \ddots & \\ & & \ddots & \lambda \\ & & & \mu \end{pmatrix}\right\} \eta + 1;$$

1980 *Mathematics Subject Classification* (1985 *Revision*). Primary 20D15, 20F05.
Translation of Dokl. Akad. Nauk Ukrain. SSR Ser. A **1980**, no. 9, 9–11; MR **82d**:20026.

$$2)\quad N_\nu = \left.\begin{pmatrix} \mu & \lambda & & & \\ & \mu & \lambda & & \\ & & \ddots & \ddots & \\ & & & \ddots & \lambda \\ & & & & \mu \end{pmatrix}\right\}\nu;$$

3) $\mu\Phi + \lambda E$, where Φ is the Frobenius cell corresponding to the degree of irreducibility of the polynomial over F and E is the identity matrix of the same size as Φ.

4) a rectangular zero matrix.

THEOREM 2. *The pencil C of matrices over F associated with G is the direct sum of matrices of types* 1)–4), *and it determines the commutation relations in G.*

PROOF. If $\{g_1, \dots, g_n\}$, $\{h_1, \dots, h_n\}$, and $\{z_1, z_2\}$ are bases of H_1, H_2, and G', respectively, then

$$[g_i, h_j] = z_1^{\alpha_{ij}} z_2^{\beta_{ij}} \tag{1}$$

where it can be assumed that α_{ij}, $\beta_{ij} \in F$ $(1 \leq i \leq n,\ 1 \leq j \leq m)$. Set $A = (\alpha_{ij})$ and $B = (\beta_{ij})$. The pencil $C = \mu A + \lambda B$ can be reduced to a direct sum of matrices of types 1)–4) by a suitable choice of bases for H_1 and H_2 (see [5], Chapter XII, §4).

THEOREM 3. *The group G decomposes into a central product of subgroups to each of which corresponds a matrix of one of the types* 1)–4). *In addition, the commutator subgroups of those factors with commutator subgroups of order p^2 are identified.*

We introduce notation for the parameters of the direct summands of C (see Theorem 2). Let t be the sum of the number of rows and columns of the zero matrix—direct sum of summands of type 4)—and let $X_1 = \{\varepsilon_1, \dots, \varepsilon_r\}$, $X_2 = \{\eta_1, \dots, \eta_s\}$, $X = \{X_1, X_2\}$, and $Y = \{g_1(\lambda, \mu), \dots, g_u(\lambda, \mu)\}$, where $g_k(\lambda, \mu)$ $(k = 1, \dots, u)$ is a homogeneous binary form which to within a constant factor is equal to the determinant of a matrix of type 2) or 3). This factor is incidental.

A simply ordered triple $R = \{t, X, Y\}$ is specified by the choice of basis for G'. Changing the basis of G' does not affect the pair t, X, but the forms $g(\lambda, \mu) \in Y$ go over into equivalent ones:

$$g(\lambda, \mu) \to g(v_1\lambda + v_2\mu, v_3\lambda + v_4\mu),$$

where $v_1, \dots, v_4 \in F$. We denote by $\overline{Y}$ the simultaneous (that is, with the same $v_1, \dots, v_4$) equivalence class of forms from Y. The class $\overline{R} = \{t, X, \overline{Y}\}$ is said to be *defining* for G. It is uniquely specified by G.

We note that if $|G'| = p$, then G may be uniquely associated with the set R which has $X = \varnothing$ and $Y = \{\lambda, \dots, \lambda\}$.

THEOREM 4. *The mapping $G \to \overline{R}$ constructed above is one-to-one.*

THEOREM 5. *To within the order of the factors, the group G can be uniquely decomposed as a central product of indecomposable factors of the same class but with commutator subgroups of orders dividing p^2. The product of all abelian factors in this decomposition is an elementary abelian p-group of rank t and splits off as a direct factor of G. The non-abelian factor H is determined either 1) by the number of $\varepsilon_i \in X_1$ or $\eta_j \in X_2$, or 2) by the equivalence class generated by the degree of irreducibility of the binary form over F.*

In the first case H is defined by $\varepsilon = \varepsilon_i$ as follows Suppose that H_1 and H_2 are elementary abelian p-groups of ranks ε and $\varepsilon + 1$, respectively, with bases $\{g_1, \dots, g_\varepsilon\}$ and $\{h_1, \dots, h_{\varepsilon+1}\}$. We construct H by taking the nilpotent product H_0 of class 2 of H_1 and H_2 and impose upon it relations of type (1), where z_1 and z_2 are arbitrary elements of a basis of H_0' and the matrices $(\alpha_{ij}) = A$ and $(\beta_{ij}) = B$ are given by the equation $\mu A + \lambda B = L_\varepsilon$. If H is associated with $\eta = \eta_j$ then H_1 and H_2 interchange ranks (that is, H_1 has a basis $\{g_1, \dots, g_\varepsilon + 1\}$ and H_2 has a basis $\{h_1, \dots, h_\varepsilon\}$), and $\mu A + \lambda B = L_{\eta'}$.

Finally, in the second case, if H is associated with an equivalence class of binary forms f, then $\operatorname{rank} H_1 = \operatorname{rank} H_2 = \deg f$ for any form $f \in \overline{f}$ and $\mu A + \lambda B$ is a matrix of type 2) or 3) corresponding to f. With a linear form (and only with such) there is associated a group H with commutator subgroup of order p.

In order to construct G as a central product we identify the commutator subgroups of the factors belonging to each of the following sets:

1) those factors having commutator subgroups of order p^2,

2) those factors which correspond to the same linear form.

BIBLIOGRAPHY

1. Heinz Liermann, *Endlichen Gruppen, deren Kommutatorgruppenordnung eine Primzahl $p \neq 2$ ist*, Schr. Math. Inst. und Inst. Angew. Math. Univ. Berlin **4** (1939), 183–207.

2. V. V. Sergeĭchuk, *Finite p-groups with commutator subgroups of order p*, Twelfth All-Union Algebra Colloq. (Sverdlovsk, 1973), Abstracts of Reports, Part I, Ural. Nauchn. Tsentr Akad. Nauk SSSR, Sverdlovsk, 1973, p. 107. (Russian)

3. ——, *The classification of metabelian p-groups*, Matrix Problems (Yu. A. Mitropol′skiĭ and L. A. Nazarova, editors), Inst. Mat. Akad. Nauk Ukrain. SSR, Kiev, 1977, pp. 150–161.

4. Rudolf Scharlau, *Paare alternierender Formen*, Math. Z. **147** (1976), 13–19.

5. F. R. Gantmakher, *The theory of matrices*, 2nd ed., "Nauka", Moscow, 1966; English transl. of 1st ed., Vols. 1,2, Chelsea, New York, 1959.

Translated by J. C. LENNOX

Amer. Math. Soc. Transl.
(2) Vol. **132**, 1986

Subgroups of the Full Linear Group over a Ring which Contain a Group of Block Triangular Matrices

UDC 519.46

N. A. VAVILOV

This note studies subgroups of the full general linear group $G = GL(n, \Lambda)$ over an associative ring Λ with identity which contain a group of block triangular matrices. The multiplicative group of units of the ring Λ is denoted by Λ^*, and d denotes the smallest natural number greater than or equal to 2 such that the conditions $SR_{d+1}(\Lambda)$ and $SR'_d(\Lambda)$ are satisfied (see [1], pp. 231–232).

An ordered system of natural numbers $\nu = (n_1, \dots, n_t)$ is called a *partition* of the number n if $n = n_1 + \cdots + n_t$. For $k = 1, \dots, t$ we put $r_k = n_1 + \cdots + n_k$. The set $\{r_1, \dots, r_t = n\}$ is denoted by J, and for $k = 1, \dots, t$ the set of indices i for which $r_{k-1} + 1 \leq i \leq r_k$ is denoted by I_k. The group $B(\nu)$ of block triangular matrices which correspond to the partition ν consists of those matrices $a = (a_{ij}) \in G$ for which $a_{ij} = 0$ when $i \in I_k$ and $j \in I_l$, $k > l$.

In [2], to every net $\sigma = (\sigma_{ij})$ of ideals in Λ of order n there is associated a net subgroup $G(\sigma)$ of G. We denote by $E(\sigma)$ the subgroup of $G(\sigma)$ generated by transvections $t_{ij}(\alpha) = e + \alpha e_{ij}$, $\alpha \in \sigma_{ij}$, $i \neq j$, where the e_{ij} are the standard unit matrices. We denote by $A(\sigma)$ the subgroup generated by $E(\sigma)$ and $B(\nu)$. The net $\sigma = (\sigma_{ij})$ is called *parabolic* if $\sigma_{ij} = \Lambda$ for all $i \leq j$.

THEOREM 1. *Let $\nu = (n_1, \dots, n_t)$ be a partition of the natural number n in which $n_k \geq d$ for all k. Then, for each subgroup H of $G = GL(n, \Lambda)$ which contains the group $B(\nu)$ of block triangular matrices, there is a unique parabolic net σ such that $A(\sigma) \leq H \leq G(\sigma)$.*

Given the group H, it is easy to construct the largest net σ such that $E(\sigma) \leq H$. The main idea of the proof of Theorem 1 consists of this: multiplying any

1980 *Mathematics Subject Classification* (1985 *Revision*). Primary 20G35.

Translation of Vestnik Leningrad. Univ. **1977**, no. 19 (Ser. Mat. Mekh. Astr. vyp. 4), 139–140; MR **57** #3278.

matrix of H by elements of $A(\sigma)$, one can obtain a matrix, a given column of which coincides with the corresponding column of the identity matrix.

The condition $n_k \geq d$ can be weakened somewhat. Namely, for one of the values of k it suffice to require that $n_k \geq 2$. If the ring Λ is commutative, then, in the case $t = 2$ or where the ideal of Λ generated by the elements $\varepsilon - 1$ for all invertible ε coincides with Λ, one of the n_k may be taken equal to 1.

For semilocal rings Λ all parabolic subgroups of $GL(n, \Lambda)$ are described in [2]; that is, subgroups containing the group of upper triangular matrices. The main results of [2] are true for rings which satisfy condition SR_2.

For Dedekind rings, the description of the subgroups H in Theorem 1 admits the following refinement. For a parabolic net $\sigma = (\sigma_{ij})$ we put $\sigma_r = \sigma_{r+1,r}$, $1 \leq r \leq n-1$, and σ_n equal to the null ideal.

THEOREM 2. *If R is a Dedekind ring, then, under the hypotheses of Theorem 1 (with $d = 2$), the subgroup $A(\sigma)$ is a normal subgroup of $G(\sigma)$ and the factor group $G(\sigma)/A(\sigma)$ is isomorphic to the direct product*

$$\prod_{r \in J} (R/\sigma_r)^* / R^* \bmod \sigma_r,$$

where $R^* \bmod \sigma_r$ denotes the subgroup of $(R/\sigma_r)^*$ consisting of those classes of $(R/\sigma_r)^*$ which have a representative in R^*.

The proof is based on the results of [3].

If it is assumed that at least two n_k of the partition $\nu = (n_1, \ldots, n_t)$ are equal to 1, then the assertion in Theorem 2 does not hold in general.

Theorems 1 and 2 are also correct for subgroups of the special linear group $\Gamma = SL(n, R)$ which contain $B(\nu) \cap \Gamma$. In the case where R is a Euclidean ring and the number t of diagonal blocks is equal to 2, analogous results were established in [4].

BIBLIOGRAPHY

1. Hyman Bass, *Algebraic K-theory*, Benjamin, 1968.

2. Z. I. Borevich, *On parabolic subgroups in linear groups over a semilocal ring*, Vestnik Leningrad. Univ. **1976**, no. 13 (Ser. Mat. Mekh. Astr. vyp. 3), 16–24; English transl. in Vestnik Leningrad Univ. Math. **9** (1981).

3. N. A. Vavilov, *Parabolic congruence subgroups in linear groups*, Zap. Nauchn. Sem. Leningrad. Otdel. Mat. Inst. Steklov. (LOMI) **64** (1976), 55–63; English transl. in J. Soviet Math. **17** (1981), no. 2.

4. N. S. Romanovskiĭ, *The subgroups of the full and the special linear group over a ring*, Mat. Zametki **9** (1971), 699–708; English transl. in Math. Notes **9** (1971).

Translated by R. SANDLING

ABCDEFGHIJ — 89876